T0134556

Studies in Computational Intelligence

Volume 796

Series editor

Janusz Kacprzyk, Polish Academy of Sciences, Warsaw, Poland
e-mail: kacprzyk@ibspan.waw.pl

The series "Studies in Computational Intelligence" (SCI) publishes new developments and advances in the various areas of computational intelligence—quickly and with a high quality. The intent is to cover the theory, applications, and design methods of computational intelligence, as embedded in the fields of engineering, computer science, physics and life sciences, as well as the methodologies behind them. The series contains monographs, lecture notes and edited volumes in computational intelligence spanning the areas of neural networks, connectionist systems, genetic algorithms, evolutionary computation, artificial intelligence, cellular automata, self-organizing systems, soft computing, fuzzy systems, and hybrid intelligent systems. Of particular value to both the contributors and the readership are the short publication timeframe and the world-wide distribution, which enable both wide and rapid dissemination of research output.

More information about this series at http://www.springer.com/series/7092

María Eugenia Cornejo · László T. Kóczy
Jesús Medina · Antonio Eduardo De Barros Ruano
Editors

Trends in Mathematics and Computational Intelligence

 Springer

Editors
María Eugenia Cornejo
Department of Mathematics,
Faculty of Engineering
University of Cádiz
Puerto Real, Cádiz, Spain

Jesús Medina
Department of Mathematics,
Science Faculty
University of Cádiz
Puerto Real, Cádiz, Spain

László T. Kóczy
Faculty of Engineering Sciences
Széchenyi István University
Győr, Hungary

Antonio Eduardo De Barros Ruano
Faculty of Science and Technology
University of Algarve
Faro, Portugal

and

Budapest University of Technology
and Economics
Budapest, Hungary

ISSN 1860-949X   ISSN 1860-9503   (electronic)
Studies in Computational Intelligence
ISBN 978-3-030-13117-3   ISBN 978-3-030-00485-9   (eBook)
https://doi.org/10.1007/978-3-030-00485-9

This Springer imprint is published by the registered company Springer Nature Switzerland AG
The registered company address is: Gewerbestrasse 11, 6330 Cham, Switzerland

Preface

Many areas of knowledge have the need to solve problems in which the use of mathematical techniques and computational intelligence is fundamental. Important research topics are focused on developing the interaction between computational intelligence and mathematics, in order to address different challenges of the current technological age.

In this volume, these two research areas are connected in appealing contributions devoted to give solutions to some interesting and real problems, and it is mainly composed of the extended and reviewed versions of the highest quality papers presented at the 9th European Symposium on Computational Intelligence and Mathematics (ESCIM 2017) held at Faro (Portugal), from October 4 to 7. Moreover, the technical program of this conference included five excellent keynote speeches, given by Profs. Jesús Medina (What is a Galois connection?), Rosana Rodríguez-López (On some properties of the solutions to fuzzy differential equations), Vera Kurkova (Capabilities of shallow and deep networks), Francisco J. Valverde-Albacete (Why are there so many semifields in computational intelligence?), and Debjani Chakraborty (Fuzzy geometry).

Next, a brief summary of the contributions contained in this volume is introduced:

The first paper employs intelligent methods for modeling of cancellous bone by using the analysis of images derived from micro-CT scans. The developed method is based on metaheuristic imitation of the flower pollination process in order to optimize the internal parameters of the Niblack binarization algorithm.

The second work addresses the problem of identifying traffic events from videos. Multi-agent systems have proven to be a successful software technology with applications in a large number of fields. For the identification of the events in traffic videos, the authors present a software architecture based on a multi-agent system which provides a significant reduction in the amount of data to be processed. This reduction makes feasible the creation of models in real time.

The goal of the third paper is to detect polyps on pictures for the correct diagnosis of possible malignant lesions. The authors introduce a fuzzy classification method to determine if a given image segment contains parts of polyps.

An application of the classification algorithm to a database of images is presented in this work.

The fourth paper presents the notion of maximal specific probabilistic law from a given probability on a set of logical formulas. A prediction operator is also defined by using maximal specific laws. The authors prove that a consistent set of consequences are obtained when the prediction operator is applied to some consistent set of logical formulas.

The fifth paper focuses on the formulation of an unfolding transformation, in order to optimize symbolic multi-adjoint logic programs and generate more efficient code. The authors show that the procedural semantics for multi-adjoint logic programs and symbolic multi-adjoint logic programs are based on a similar scheme. Specifically, the interpretive stage always provides a value for multi-adjoint logic programs, while for symbolic multi-adjoint logic programs, the expressions containing symbolic values and connectives must be instantiated in order to obtain a final value.

The sixth paper introduces an interesting mechanism to represent multi-adjoint logic programs by using hypergraphs and their corresponding subjacent directed graphs. This representation allows a generalization of a termination theorem, which plays an important role in the semantics of the multi-adjoint logic programming framework.

In the seventh paper, the authors model the internal operation of a department of a bank by using Fuzzy Cognitive Map, which is a suitable tool to simulate the behavior of complex systems. They introduce a method which is capable of modifying the existing connections in the Fuzzy Cognitive Map and testing the effects of these changes. This fact provides an important advantage; the hidden behavior of the model as well as the concepts with a greater degree of influence can be detected.

The aim of the eighth paper is to provide a contextual reasoner capable of adapting the big knowledge basis into affordable problems. To reach this goal, the authors use the language Scala, to implement an inference rule which allows us to carry out a reduction in the size of the knowledge basis based on the retraction problem.

The ninth paper is located in the framework of Formal Concept Analysis. Usually databases contain an absence of information which is often ignored, and it can provide relevant information. The authors of this work address this problem and propose a new logic to build inference rules to handle with positive and negative information.

In the tenth paper, the authors study the relationship between two important theories to analyze and process information contained in databases: Rough Set Theory and Formal Concept Analysis. Specifically, they analyze the existing connections from the perspective of attribute reduction. They establish a sufficient and necessary condition in order to ensure that reducts in both theories coincide.

The eleventh paper proposes a new similarity measure between linguistic terms based on restricted equivalence functions. This novel similarity measure is implemented and incorporated in the core of the Bousi Prolog system, which allows an enhancement of the Bousi Prolog system inference engine. An experimental comparison among similarity measures is also carried out.

The twelfth paper deals with two problems related to the online adaptation scheme: parameter shadowing and parameter interference. To solve these problems, the authors propose a new online adaptation method based on the convex hull concept. In addition, in order to update the fixed-structure Radial Basis Function Neural Networks models, a sliding-window technique is introduced.

In the thirteenth work, the authors present an evolutionary metaheuristic in order to solve the minimum latency problem. An algorithm is introduced, which is proved to be efficient since the accuracy of the results is high and the average runtime has been smaller than other existing methods to solve the same problem.

The fourteenth work introduces a study about the search for solutions of multi-objective fuzzy geometric programming problem. Specifically, the authors introduce a new methodology in order to obtain fuzzy Pareto optimal frontier. The presented mechanism has the advantage that it is efficient regardless of the degree of difficulty of the problem under consideration. To solve this optimization problem, fuzzy geometry has been considered. In addition, the results presented in this work have been backed up by examples.

In the fifteenth work, the authors build a database from the information provided by a location-based social network. This database is susceptible to be treated by formal concept analysis in order to know the habits and interests of the users, which leads to build a consistent recommendation system.

The sixteenth paper introduces the notion of lattice-valued Boolean algebra together with its most interesting properties and its relationship with the notion of a generalized lattice valued lattice. The paper contains basic structures for developing generalized Boolean functions.

The seventeenth paper studies the solvability of the max-product fuzzy relation equations in which a negation operator is considered. Specifically, the residuated negation of the product t-norm is introduced in these equations. The solvability and the set of solutions of these bipolar equations have been studied in different scenarios, depending on the considered number of variables and equations.

The eighteenth paper carries out a comparison between direct and indirect methods for solving two-mode systems of fuzzy relation equations and inequalities. Interesting technical results prove that the solutions of any two-mode systems of fuzzy relation equations and inequalities can be obtained from the solutions corresponding to the related one-mode system, and vice versa.

The nineteenth paper extends the notion of context and the definition of concept-forming operators given in the formal concept analysis framework to the intuitionistic L-fuzzy sets setting. This extension arises as a consequence of the construction of an adjoint pair for intuitionistic L-fuzzy values.

In the twentieth paper, the authors present a study about the dependence between random variables. They introduce new measures of independence within the possibility theory. Specifically, they define the possibilistic versions of the distance covariance and correlation given in probability theory and statistics.

The twenty-first paper presents preliminary results concerning a fuzzy approach for measuring sentence checkability. The proposed fuzzy approach has been implemented and tested in a large range of sentences, obtaining sound and optimistic results. Moreover, the authors carry out a comparison between the obtained results and the ClaimBuster score, which gives rise to a favorable validation.

The twenty-second paper develops a practical and integrated methodology in order to offer support for the promotion of Circular Economy. The developed methodology considers reduced model investigations, which are supported by Fuzzy Cognitive Maps, for evaluating local and regional opportunities involved in the realization of Circular Economy.

Finally, we would like to finish this preface showing our acknowledgment to the authors, since without their effort and interest this special issue would not have been possible. A word of thanks is also due to EasyChair, for the facilities provided in the submission/acceptance of the papers and in the preparation of this book.

Puerto Real, Spain   María Eugenia Cornejo
Győr, Hungary   László T. Kóczy
Puerto Real, Spain   Jesús Medina
May 2018

Contents

Application of the Flower Pollination Algorithm in the Analysis of Micro-CT Scans

Piotr A. Kowalski, Jakub Kamiński, Szymon Łukasik,
Joanna Świebocka-Więk, Dominika Gołuńska, Jacek Tarasiuk
and Piotr Kulczycki

Abstract The aim of this article is to present research involving the employment of intelligent methods for image analysis, particularly, the binarization process. In this case, the Flower Pollination Algorithm was used to optimize the internal parameters of the Niblack binarization algorithm. As a criterion for the quality of the proposed solution, the morphological parameter called the 'bone volume' (equivalent to porosity) is taken into account. The overarching objective of this study is to model the structure of cancellous bone based on the analysis of images derived from Micro-CT Scans.

P. A. Kowalski (✉) · J. Kamiński · S. Łukasik · J. Świebocka-Więk · D. Gołuńska · J. Tarasiuk
P. Kulczycki
Faculty of Physics and Applied Computer Science AGH, University of Science and Technology,
al. Mickiewicza 30, 30-059 Cracow, Poland
e-mail: pkowal@agh.edu.pl; pakowal@ibspan.waw.pl

J. Kamiński
e-mail: kaminski@fis.agh.edu.pl

S. Łukasik
e-mail: slukasik@agh.edu.pl; slukasik@ibspan.waw.pl

J. Świebocka-Więk
e-mail: jsw@agh.edu.pl

D. Gołuńska
e-mail: golunska@agh.edu.pl

J. Tarasiuk
e-mail: tarasiuk@agh.edu.pl

P. Kulczycki
e-mail: kulczycki@agh.edu.pl; kulczycki@ibspan.waw.pl

P. A. Kowalski · S. Łukasik · P. Kulczycki
Systems Research Institute, Polish Academy of Sciences,
ul. Newelska 6, PL-01-447 Warsaw, Poland

© Springer Nature Switzerland AG 2019
M. E. Cornejo et al. (eds.), *Trends in Mathematics and Computational
Intelligence*, Studies in Computational Intelligence 796,
https://doi.org/10.1007/978-3-030-00485-9_1

Keywords Computational intelligence · Flower pollination algorithm
Biologically inspired algorithm · Micro-CT · Binarization · Scans processing

1 Introduction

In recent times, computational intelligence (CI) methods have been increasingly
applied in science and technology or economics. Herein, biologically inspired optimization methods are one of the major CI groups inclusive of many procedures
based on diverse mechanisms of search for suboptimal solution [9], and numerous
applications of optimization algorithms base on CI in wide areas [4, 7, 15] can be
underlined. In this study, we will use the Flower Pollination Algorithm (FPA) to
optimize the parameters of the image analysis algorithm.

Medical image analysis is a most dynamically developing form of explorative data
analysis. In the case of 3D imaging, the primary purpose of describing a material's
structure is usually to reveal certain characteristic features (i.e. shape and placement)
of its construction. The integrity and differentiation of bone microstructure affects its
mechanical properties, and microstructure evaluation may be useful in both fracture
and pathological evaluation, including that osteoporosis-induced [1]. The average
morphometry characteristics of structure and porosity are determined taking into
account the specific volume within the analyzed sample. In the analysis of this type
of tissue, the complexity of the microstructure is reflected only in the sense of average
value. Therefore, experimental and theoretical models of their properties refer to the
microstructural features of the medium averaged in a given volume, and, should,
hence, be interpreted [18].

The bovine femur bone (Fig. 1) used in this study corresponds to the intended
purpose. It is characterized by high variability in the microstructure, especially in
the case of volume fractions, and, thus, porosity [5].

Fig. 1 Bovine cancellous
bone, along with the tested
cuboid segment

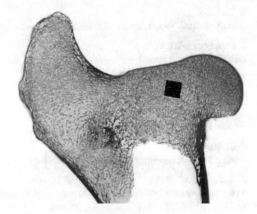

Among all the non-contact methods, computed microtomography (micro-CT) is the most successful in the context of cancellous bone microstructure research. This is a non-invasive test method which allows a reproduction of the internal microstructure of the examined object on the basis of the two-dimensional projections recorded at different angles. Micro-CT builds upon the same assumptions as classical computer tomography, but by using a smaller dot focusing on the electron beam, it is possible to obtain a higher resolution of the reconstructed image. Due to its relatively high resolution and non-destructive nature, it has been used in many fields of science and technology, but its use in medicine (in the imaging of tissues and organs) has led to its intensive development [17]. Currently, stationary devices of this type are able to achieve sub-micrometre resolution, but the implementation of this technology involved overcoming two major obstacles. The first was the need to use computers capable of processing large amounts of data in as short a time as possible. Equipment of this kind was often unreachable for financial reasons. The second, more important obstacle was the need for a high-resolution X-ray system with a sub-focal spot lamp and high enough power to allow penetration into a dense sample and to return the image at the correct resolution [2].

One of the critical algorithm used in modeling bones is image binarization. In Fig. 2—left side, we can see a bone scan generated through micro-CT technology. On the right side of this image a histogram through which we can easily determine the threshold value for the binarization procedure was presented. By way of the express separation of white from dark pixels, this parameter is global and the sense of its choice is not a key issue for the final results.

In the low-resolution imagery (see Fig. 3) obtained through normal tomography, the histogram does not have such a clear region of separation of individual pixels. Therefore, due to the nature of the examined image subject (bones), the binarization procedure must be treated as a local character algorithm. In view of the above, it is desirable to determine the preferred parameters of the binarization algorithm. In the case of the presented studies, an CI algorithm, in particular, the FPA was used to

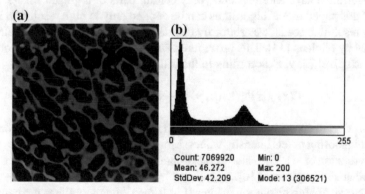

Fig. 2 High-resolution grey scale images (**a**) after registration with the information seen in the histogram (**b**)

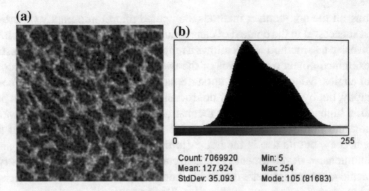

Fig. 3 Low-resolution grey scale image (**a**) after registration with the information seen in the histogram

determine the suboptimal parameters of the binarization procedure. The correctness of the solution is obtainable through combining high and low resolution imagery. The presented binarization algorithm (along with the internal parameters) will be applied for low resolution tomography scans. Hence, the low resolution tomography scan results—common in hospital test procedures—can be enhanced.

2 Methods and Algorithms

2.1 Binarization Procedure

In the image analysis domain, there are many known binarization algorithms. Among these, global algorithms have a positive overall effect upon the entire image, while local algorithms have an effect upon only certain parts of a greater image. For the topic of this paper, global algorithms can be applied only to high resolution images. For low resolution scans, we arbitrarily decided to utilize the binarization algorithm proposed by Niblack [14]. This procedure is based on calculating a threshold value T for each pixel (x, y, z) according to the following formula:

$$T(x, y, z) = m(x, y, z) + k \cdot std(x, y, z) - l \qquad (1)$$

where $m(x, y, z)$ constitutes a local mean, and $s(x, y, z)$ denotes a standard deviation of the neighboring pixels intensity values. Moreover, a k and l are weight parameters for each fraction of (1). The value of these parameters greatly influences the outcome of the binarization process [16].

In the case of high resolution images (Fig. 2-*left*), after applying a median filter, a global threshold for the Otsu binarization operation can be automatically achieved. As the end effect of this procedure, a threshold parameter O_{Otsu} is obtained. To generate binarization of the scan, an Auto Local Threshold plugin (implemented via ImageJ software [13]) is then applied. In all above cases, a binarization procedure for three-dimensional space is computed.

2.2 Flower Pollination Algorithm

FPA belongs to the class of global optimization procedures. This method is inspired by the process of flower pollination, and was introduced by Yang in 2012 [19]. This procedure (see: Algorithm 1) starts with an initialization phase. In the main loop of the algorithm, a random value is then generated which determines if global pollination or local pollination is being carried out. The first procedure is inspired by the movement of insects flying over long distances so as to achieve the act of distant pollination. This step of the heuristic algorithm corresponds to the so-called 'exploration of the solution space'. For FPA, Levy flight distribution is employed to

---

Algorithm 1 Flower Pollination Algorithm [11]

1: Initialize algorithm $k \leftarrow 1$, $f(x^*(0)) \leftarrow \infty$
2: **for** $i = 1$ to M {each individual in swarm} **do**
3:     Generate Solution $(x_i(k))$
4:     {evaluate and update best solutions}
5:     $K(x_i(0)) \leftarrow$ Evaluate quality$(x_i(0))$
6: **end for**
7: $x^* \leftarrow$ Save best individual $x^*(0)$
8: {main loop}
9: **repeat**
10:     **for** $i = 1$ to M **do**
11:         **if** $Real\_Rand\_in\_(0, 1) < prob$ **then**
12:             {Global pollination}
13:             $s \leftarrow Levy(s_0, \gamma)$ and $x_{trial} \leftarrow x_p(k) + s(x^*(k) - x_p(k))$
14:         **else**
15:             {Local pollination}
16:             $\epsilon \leftarrow Real\_Rand\_in\_(0, 1)$ and
17:             $r, q \leftarrow Integer\_Rand\_in(1, M)$
18:             $x_{trial} \leftarrow x_p(k) + \epsilon(x_q(k) - x_r(k))$
19:         **end if**
20:         $f(x_{trial}) \leftarrow$ Evaluate_quality(x_{trial})
21:         **if** Check if new solution better **then**
22:             $x^* \leftarrow$ Save best individual $x^*(k)$
23:             $K^* \leftarrow$ Save cost function value for the best individual $K(x^*(k))$
24:         **end if**
25:     **end for**
26:     find best and copy population
27:     $stop\_condition \leftarrow Check\_stop\_condition()$
28:     $k \leftarrow k + 1$
29: **until** $stop\_condition =$ **false**
30: **return** $K(x^*(k)), x^*(k), k$

---

realize this behaviour. The second process is inspired by a different, local pollination process that is commonly referred to as 'self-pollination'. Herein, a local search that leads towards the exploitation of the solution space, is implemented. To determine the ratio of global to local search processes, a parameter called 'switching probability' is put into place [11]. The best value of cost function and the argument for which it was obtained are returned as the outcome of the application of the algorithm. More details about this metaheuristic algorithm can be found in [11, 19].

3 Numerical Studies

During the study, the FPA procedure was used to determine the parameters k and l in the Niblack binarization algorithm. Based on prior studies [8, 11, 12], the following FPA procedural parameters were determined: number of swarm members 10, switching probability 0.8, maximal number of iterations 30. In each iteration of the optimization procedure, two-dimensional vectors of the solution space $[k, l]$ were evaluated using the following cost function:

$$J(k, l) = \left| \frac{BV(k, l)}{TV} - B_{Otsu} \right|. \tag{2}$$

where $BV(k, l)$ denotes the number of white pixels based on the Niblack procedure, TV constitutes the number of all pixels and B_{Otsu} indicates a reference value obtained by means of the Otsu algorithm as applied for high resolution CT scans. In this simulation, the following constraints: $0 \leq k \leq 2$ and $0 \leq l \leq 2$ are utilized.

With regard to the aforementioned criterion for the quality of the proposed solution, the morphological parameter called the 'bone volume' $\frac{BV}{TV}$ (identical to porosity) is taken into account. Furthermore, bone volume, as achieved from low resolution images, is compared with the result based on the global threshold O_{Otsu}. In both cases of resolution, in the first stage, a median filter must be applied. The difficulty of the above tests is that such are quite expensive (in the sense of computation time) when generating the evaluation function (2). Indeed, in every case of use, full binarization of the whole bone is necessary.

In Fig. 4, an approximation of the $J(k, l)$ function can be observed. Here, we can see its multimodal character, hence the application of a metaheuristic procedure for searching minimal value is completely justified. As a result of this investigation, we can distinguish two groups of solution vectors—both related to minima of (1) function. The first one (and the best) is [0, 0] with $J = 0.0670$, the subsequent group consists of the following solutions [0.4534, 1.2595], [0.0076, 1.2922], [0.3081, 1.4653], [0, 0.8736] and [0, 1.3549], for which the value of cost function is $J = 0.0760$. Here, we must underline that the first obtained solution on boundary condition is located.

In addition, in Fig. 5, the results obtained for other heuristic optimization algorithm namely 'Particle Swarm Optimization' (PSO) [3, 6] are presented. Similarly to the

FPA, the PSO procedure belongs to a group of algorithms based on the idea of swarm intelligence [20]. In this study, we used the PSO algorithm according to the recommended internal parameters described in the article [10]. In Fig. 5, using a blue (dotted) line, the convergence of the proposed procedure is presented, while the red continuous line represents the convergence of the PSO procedure. The comparison clearly indicates the significant advantage of the proposed solution to the alternative one. The FPA algorithm has a much faster convergence, and, most importantly, it achieved a better final solution.

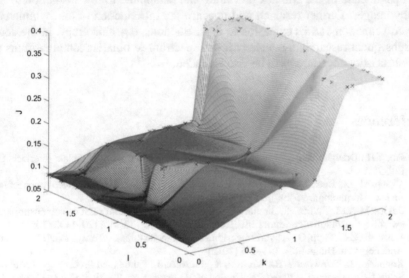

Fig. 4 Approximation of cost function based on points received during the FPA simulation

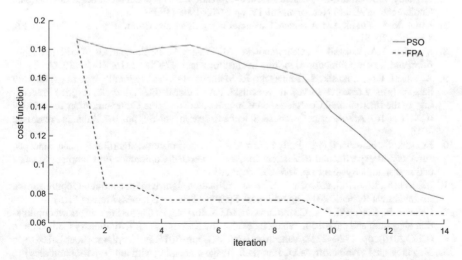

Fig. 5 Convergence comparison of FPA with respect to the PSO algorithm

4 Conclusions

In the present study, an early stage research results related to bone modelling, have been demonstrated. Developed solution is based on metaheuristic imitating the process of pollinating plants—used to determine the parameters of the binarization procedure. In this case, a simple evaluation function based on a bone volume approach was employed. However, it should be emphasized that this bone feature is one of the most important in modelling reconstruction. By way of the presented research, the best parameters of the Niblack procedure can be applied to low resolution tomography images. Further research will concern the introduction of an asymmetry of the cost function and its extension to other fractions, e.g. anisotropy. In the course of subsequent research, the already broad spectrum of binarization algorithms and metaheuristic procedures will be also extended.

References

1. An, Y.H., Draughn, R.A.: Mechanical Testing of Bone and the Bone-Implant Interface. CRC (1999)
2. Chappard, D., Basl, M.-F., Legrand, E., Audran, M.: Trabecular bone microarchitecture: a review. Morphologie **92**(299), 162–170 (2008)
3. de Oca, M.A.M., Stutzle, T., Birattari, M., Dorigo, M.: Frankenstein's PSO: a composite particle swarm optimization algorithm. IEEE Trans. Evol. Comput. **13**(5), 1120–1132 (2009)
4. Johanyák, Z.C., Papp, O.: A hybrid algorithm for parameter tuning in fuzzy model identification. Acta Polytech. Hung. **9**(6), 153–165 (2012)
5. Kamiński, J., Trzewiczek, B., Wroński, S., Tarasiuk, J.: Automated Processing of Micro-CT Scans Using Descriptor-Based Registration of 3D Images, pp. 73–79. Springer, Cham (2017)
6. Kennedy, J., Eberhart, R.: Particle swarm optimization. In: Proceedings of IEEE International Conference on Neural Networks, vol. IV, pp. 1942–1948 (1995)
7. Kıran, M.S., Fındık, O.: A directed artificial bee colony algorithm. Appl. Soft Comput. **26**, 454–462 (2015)
8. Kowalski, P.A., Łukasik, S., Charytanowicz, M., Kulczycki, P.: Comparison of krill herd algorithm and flower pollination algorithm in clustering task. ESCIM **2016**, 31–36 (2016)
9. Kowalski, P.A., Łukasik, S., Kulczycki, P.: Methods of collective intelligence in exploratory data analysis: a research survey. In: Kowalski, P.A., Łukasik, S., Kulczycki, P. (eds.) Proceedings of the International Conference on Computer Networks and Communication Technology (CNCT 2016). Advances in Computer Science Research, vol. 54, pp. 1–7. Atlantis Press, Dec 2016
10. Łukasik, S., Kowalski, P.A.: Fully informed swarm optimization algorithms: basic concepts, variants and experimental evaluation. In: 2014 Federated Conference on Computer Science and Information Systems, pp. 155–161, Sept 2014
11. Łukasik, S., Kowalski, P.A.: Study of flower pollination algorithm for continuous optimization. In: Intelligent Systems 2014, pp. 451–459. Springer Science Business Media (2015)
12. Łukasik, S., Kowalski, P.A., Charytanowicz, M., Kulczycki, P.: Clustering using flower pollination algorithm and calinski-harabasz index. In: IEEE Congress on Evolutionary Computation (CEC 2016), pp. 2724–2728. Vancouver (Canada), July 2016. Proceedings: paper E-16413
13. Magalhaes, P.J., Abramoff, M.D., Ram, S.J.: Image processing with image-j. Biophotonics Int. **11**(7), 36–42 (2004)

14. Niblack, W.: An Introduction to Digital Image Processing. Strandberg Publishing Company, Birkeroed, Denmark (1985)
15. Precup, R.-E., David, R.-C., Petriu, E.M., Preitl, S., Rădac, M.-B.: Novel adaptive charged system search algorithm for optimal tuning of fuzzy controllers. Expert Syst. Appl. **41**(4), 1168–1175 (2014)
16. Samorodova, O.A., Samorodov, A.V.: Fast implementation of the niblack binarization algorithm for microscope image segmentation. Pattern Recogn. Image Anal. **26**(3), 548–551 (2016)
17. Sensen, C.W., Hallgrímsson, B.: Advanced Imaging in Biology and Medicine: Technology, Software Environments, Applications. Springer (2008)
18. Stock, S.R.: MicroComputed Tomography: Methodology and Applications. CRC Press (2008)
19. Yang, X.-S.: Flower pollination algorithm for global optimization. Lecture Notes in Computer Science, vol. 7445, pp. 240–249 (2012)
20. Yang, X.S.: Nature-Inspired Metaheuristic Algorithms. Luniver Press, Frome (2008)

Enhancing the Analysis of Video Time Series by Means of a Multi-agent Architecture

L. Rodriguez-Benitez, J. Giralt, D. Merino, L. Jimenez-Linares and J. Moreno-Garcia

Abstract In this paper we present a software architecture based on a multi-agent system whose major goal is the identification of traffic events from videos. In order to achieve this, H264/AVC motion vectors that appear in compressed video signal are taken as input. They are classified depending on their position in the scene and after that each group of motion vectors obtained from such classification is processed independently using statistical techniques. The use of this kind of techniques have been broadly used in the processing of time series like the one we take as input. After the statistical processing, individual results are compared between them in order to detect patterns related to possible traffic events. This comparison process can be understood as a cooperative process. So, to integrate the different processing components of this architecture we propose the use of a multi-agent system. Multi-agent systems allows to define a cooperative architecture using individual agents that can be run in parallel allowing to raise the performance and efficiency of the global process of event identification. The experimentation of this paper is driven to the detection of objects in complex traffic scenarios where the videos are captured from on-board cameras.

Supported by the project TIN2015-64776-C3-3-R of the Science and Innovation Ministry of Spain, co-funded by the European Regional Development Fund (ERDF).

L. Rodriguez-Benitez (✉) · J. Giralt · D. Merino
ORETO Research Group, Escuela Superior de Informatica, University of Castilla-La Mancha, Paseo de la Universidad, 4, 13071 Ciudad Real, Spain
e-mail: luis.rodriguez@uclm.es
URL: http://oreto.esi.uclm.es

J. Giralt
e-mail: juan.giralt@uclm.es

D. Merino
e-mail: dario.merino@uclm.es

L. Jimenez-Linares
Escuela de Ingenieria Industrial, University of Castilla-La Mancha, Avd. Carlos III, s/n, 45071 Toledo, Spain
e-mail: luis.jimenez@uclm.es

© Springer Nature Switzerland AG 2019
M. E. Cornejo et al. (eds.), *Trends in Mathematics and Computational Intelligence*, Studies in Computational Intelligence 796,
https://doi.org/10.1007/978-3-030-00485-9_2

Keywords Time series · Compressed domain video · Multi-agent systems

1 Introduction

Nowadays in our society the main mean of transport for goods and people is the road transport. So, the improvement of safety and security in public transportation by means of the use of new technologies can be considered as a major topic of interest. The use of such technologies has reduced the number of traffic incidents in a direct way and that is why new efforts in the development of novel solutions are needed. On the other hand, a relevant field of research is the one related to the statistical analysis of time series. A video can be considered as a time series once we have a set of data ordered temporally even more when we consider only raw data from motion vectors in the compressed domain.

The development of new technologies can be focused from different points of view, for example, the detection in real time of vehicles, obstacles or dangerous situations. Another techniques analyze situations after they occur with a different number of exhaustive methods allowing to prevent similar situations in a future. The way these two different situations are treated is completely different with respect to the time needed to find a response and the computational resources, among others.

In this paper we present the design of a forensic solution for video analysis captured from a car in motion. The goal of this method in a first stage is the segmentation of the different objects in the scene and after that the identification of events by means of the processing of the segmentation results. For the identification of events we propose the use of a multi-agent platform as we design this identification process as a cooperative process where different computing units compare the information retrieved from the video signal. Multi-agent systems [2, 3, 6] can be considered as a software technology with successful applications in a wide range of fields like electronic commerce, robotics, information retrieval, etc.

With respect to the processing of time series in the literature it can be found a large number of methods that process them using statistical methods. The statistical methods used are very different. For example, in [8] intrinsic mode functions are used to address the problem of trend filtering. The proposed method is tested in real-world data collected from an environmental study and from a bicycle rental service. Huang and Schmitt [5] use empirical mode decomposition based on time dependent intrinsic correlation to treat time series of temperature and dissolved oxygen time series. In [10] they work on accurate estimation of long-term sea conditions that is important for the design of coastal and offshore structures, for the preparation of marine operations, and for other applications. The authors present a study on swell at a West Africa location. They use a time-consistent triangular model for the spectral shapes of the swell components, then perform a statistical analysis of the time-histories of those components in connection with the storms at their source. Lorentzen [7] studies the sea temperature data sampled at Station-M in the Norwegian Sea in the period 1948–2010. He analyzes issues such as the stochastic process that characterize time series, patterns that indicate climate change, characteristics in the data associated with ice decreasing, or whether series can be modeled to predict future temperatures. It uses different methods, such as Augmented Dickey-Fuller tests, ARIMA-models, cointegration and error-correcting models, Granger-causality tests, and simultaneous equation systems. Mueen et al. [9] study how to join two long time series based on their most correlated segments, where they can be joined at any locations and for arbitrary length. This study can be applied in different domains such as environmental monitoring, patient monitoring and power monitoring.

This paper is structured as follows. The input information is directly extracted from the video compressed domain, then in Sect. 2 information related to the H264/AVC standard is presented paying special attention to the motion compensation techniques that produces the motion vectors. Furthermore, in this section we present the statistical techniques used to process the information from the video. The implementation of such operations are embedded in some of the agents of the global architecture of agents. After that the design of the multi-agents platform is detailed in Sect. 3. Finally, the experimentation and the conclusions are shown in Sects. 4 and 5, respectively.

2   Basic Processing Information

In this section we present the information to be taken and how exactly is processed by using statistical techniques in order to obtain relevant information that could be the adequate to use in later stages to identify events. More concretely, in Sect. 2.1 relevant aspects about the video time series taken as input are detailed whilst in Sect. 2.2 a statistical method for processing such time series is proposed.

2.1   *Video Image Processing*

H.264 or MPEG4-AVC is a standard for the definition of a high quality and high compression rate video. It was developed by the ITU-T Video Coding Experts Group(VCEG) and the ISO/IEC MPEG. Basically, this standard reduces the amount of data needed to reproduce high quality video. The encoders process each frame or picture composing the whole video. The image is divided in partitions called blocks and for each block a similar one in a previous or future frame is searched. This technique is known as motion compensation or motion estimation. Once a similar block is found in another frame it is not coded, only the displacement of the block between frames is stored as a motion vector.

Analyzing traffic videos captured from a car in motion is a complex task because there exists different motion patterns depending on the area of the picture. So, a global analysis of the picture can produce confusing results. In order to process independently each one of the flow patterns that could be present in a traffic video, we propose a division of the picture. We establish next classification of areas:

South, corresponding to the dashboard and the front part of the car, Center, usually corresponds with the road, West and East corresponding with the edges of the road and/or the hard shoulders and finally, the North where cars on opposite lane or the horizon appears. For example, for a video resolution of 640X480 a possible division specifying coordinates X and Y taken (0,0) as the top left corner is shown in Fig. 1 and the values for the coordinates are:

- **North**: $y \in [0, 175]$
- **South**: $y \in [375, 480]$
- **West**: $x \in [0, 120]$; $y \in (175, 375)$
- **East**: $x \in [400, 640]$; $y \in (175, 375)$
- **Center**: $x \in (120, 400)$; $y \in (175, 375)$

As the South usually corresponds to the front part of the car no valid information for the detection of events can be extracted. So, the data from this area is not considered for future processing stages.

Fig. 1 Division of the picture in areas

2.2   Statistical Extraction of Information from Video Time Series

There exists a group of agents dedicated to the statistical processing of information obtained from H264/AVC motion vectors. In this section we detailed our proposal for such statistical analysis. The identification of motion patterns is directly related to the appearance and disappearance of fields of motion vectors. For instance, as it is shown in Fig. 2a the existence of an event as a vehicle approaching in the opposite direction produces a field of motion vectors that does not exist if there is no vehicle in the scene, Fig. 2b.

Anyway, the motion vectors are generated by internal encoding procedures that produces for example that in some frames no motion vector appears, in another ones although there exists field of motion vectors they cannot be associated to any object or motion pattern in the scene. Then the study of the behavior of the vectors cannot be restricted to the analysis of a frame. So the evolution of the motion information in a set of neighbor frames must be completed in order to generate any reliable result. Now, with the previous premise, we present our statistical method of processing.

Let $F_i(A)$ be the number of motion vectors, detected in the area A in the frame number i of the video, where A can take any of the values in the set {North, Center, East, West} as detailed in Sect. 2.1. Then in order to detect any kind of event, m consecutive frames must be analyzed, from $F_i(A)$ to $F_{i+m+1}(A)$. The basic idea of the method is to detect the appearance of fields of motion vectors associated to the appearance of objects. Then it is established a comparison between the number of motion vectors in a frame and the arithmetic mean of the number of motion vectors in m subsequent frames, always done for each one of the division areas independently. Equation 1 shows how to compute this mean.

$$M_i(A) \leftarrow \frac{\sum_{j=i+1}^{j=i+m+1} F_j(A)}{m} \qquad (1)$$

(a)   **(b)**

Fig. 2 Comparison of vectors in a frame with an object and another with no one

To establish a comparison we propose the study of the evolution of the number of motion vectors by using statistical measures such as the Standard Deviation ($S_i(A)$) (Eq. 2) for the m subsequent frames of the current one, i.

$$S_i(A) \leftarrow \sqrt{\frac{\sum_{j=i+1}^{j=i+m+1}\left(F_j(A) - M_i(A)\right)^2}{m}}$$ (2)

By means of the confidence interval significant differences between the number of motion vectors and the arithmetic mean of this number in subsequent frames can be detected. For the computation of the Confidence Interval a Student's t distribution is used. Basically, with this distribution we can estimate the mean of a population once it is assumed this mean is not normally distributed. Once the deviation standard is computed the amplitude of the confidence interval is computed in Eq. 3 where $m - 1$ are the degrees of freedom and α determines the confidence level. For example, to get a confidence level of 99% α is equal to 0.01.

$$Amp(M_i(A)) \leftarrow t_{m-1;\frac{\alpha}{2}} \times \frac{S_i(A)}{\sqrt{m}}$$ (3)

Once the amplitude is computed the confidence interval is obtained by means of Eq. 4.

$$CI_i(A) = [M_i(A) - Amp(M_i(A)), M_i(A) + Amp(M_i(A))]$$ (4)

To conclude, it must be stated that the way to determine the presence of an event in a concrete frame is the existence of a number of vectors in a concrete area out of the bounds of the confidence interval for such frame.

3 Design of the Multi-agent Architecture

Now the design of the agents architecture is introduced. Such design is determined by the development technology selected, JADE [1], a framework implemented in JAVA fulfilling the FIPA standard. FIPA is an organization for the definition of standards promoting technologies related to agents and the interoperability between agent platforms developed using different technologies. In order to establish the communication for the agents the Agent Communication Language (FIPA-ACL) [4] from FIPA has been selected. In FIPA-ACL a message contains next fields: sender, receiver, content, language, performative (communicative act [11]) and protocol, among others. Usually, the content, sender and receiver change from one message to another. Nevertheless, protocols provide of templates that can be used when the communicative intention is similar.

The design of the agents platform must be driven to obtain results in an efficient and reliable way. So, the most relevant decisions to be taken in such design are related

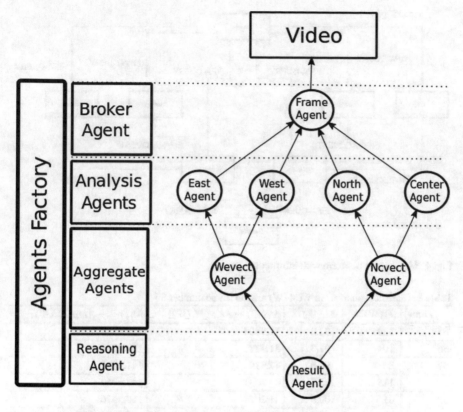

Fig. 3 Configuration of the agents platform for a specific configuration

to the number of agents in the system, their functionalities and the relationships among them. For example, these decisions directly affect to the desire achievement of a high degree of parallelism and a correctly balanced processing load. The design of the relationships determines the communication flows between agents and the intentionality of such communicative act determines the protocol to be used. The proposed factory of agents is composed of four different kind of agents:

- **Broker Agent**: This agent is the responsible of reading the configuration file for every different video analyzed. The configuration information is read and processed and after that a concrete instance of the platform of agents. So the number of other kind of agents, for instance, depend on such configuration.
- **Analysis Agent**: It receives the information about the division of the picture that corresponds to one of the areas. With these coordinates the agent extracts the information related to the motion vectors in its corresponding area.
- **Aggregation Agent**: The information about the motion vectors is received and then the statistical measures defined in Sect. 2.2 are computed. This agent is able to process the statistical data for two areas in parallel.

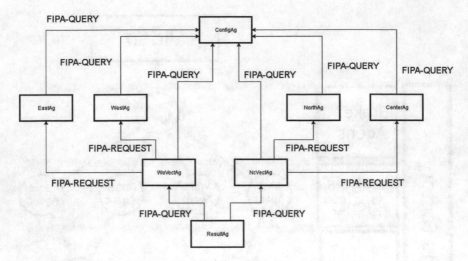

Fig. 4 Multi-agent platform with information flow

Table 1 Statistical analysis in West (W) area in frame number 35

$i(frame)$	$F_i(W)$	$M_i(W)$	$M_i(W) - Amp(M_i(W))$	$M_i(W) + Amp(M_i(W))$
35	**139**	93.0	**51.9804**	**134.0196**
36	104	91.3	51.6222	130.9778
37	86	94.6	47.9791	141.2209
38	115	95.7	43.7031	147.6969
39	81	100.9	41.2024	160.5976
40	107	98.8	37.9925	159.6075
41	98	101.6	35.8298	167.3702
42	82	104.9	41.4358	168.3642
43	103	106.2	41.9713	170.4287
44	80	112.7	45.6012	179.7988
45	74	120.6	57.1246	184.0754
46	87	127.9	65.2499	190.5501

- **Reasoning Agent**: In this agent a criterion about the relationship between results of the statistical processing and events is established. In this case, events are related to the limits of the confidence interval as shown in Eq. 3. The output is a sequence of frames where an event occurs.

A concrete configuration of the platform, with four analysis agents and two aggregation agents, is shown in Fig. 3 whilst Fig. 4 shows the information flow for this concrete execution. The design of the proposed architecture comes determined by several factors like the improvement of the parallelization in the mathematical operations and the execution of the behaviors. Another important factor is related to

Fig. 5 Frames 33, 35, 38 and 44

communications. They must be as reduced as possible and the use of a blocking and nonblocking semantic in send and receive primitives allows to control the execution flow. So, although the execution of the agents begin simultaneously they are blocked until a concrete agent sends the data they need to operate.

4 Experimental Results

Now, we present a brief review of the full set of experiments. With the experimentation we try to demonstrate that this proposal allows the detection of events. With this aim, we have extracted manually from a video these frames where an event happens. Different tests have been completed modifying the configuration variables in equations defined in Sect. 2.2.

For instance, using a value of $m = 10$ and $\alpha = 0.01$ an event has been detected in the division area of the picture named as *West* in the frame number 35. For this frame and its m subsequent frames the resulting values for Eqs. 1–4 are shown in Table 1.

Then, the number of vectors in the frame 35, $F_{35}(W)$ is 139, greater than the upper bound of the interval, 134.0196. So, an object should appear associated to the West in this frame disappearing from this zone in the subsequent 10 frames, the value of m. In Fig. 5 frames 33, 35, 38 and 44 are shown from the analyzed video. It can be observed how there is a truck driving in the opposite direction and more concretely in the frame number 35 is when this truck enters in the West. In the frame number 44 the truck disappears from this concrete area of the picture.

5   Conclusion

In this paper, the development of a system to analyze traffic videos to detect events is presented. Once detected the idea is to design models for the prediction of these events in architectures of agents that could allow the real time processing of video information. The video processing using only information from the compressed domain reduces dramatically the amount of data to be processed, then, the design of models for real time applications is feasible. Furthermore, the use of high-level software architectures like multi-agent systems allow to encapsulate the information in a direct way completing the different computing tasks in parallel and synchronizing the global operation of the system.

As future work, it could be incorporated a new kind of agent containing the full catalog of events to be detected. Even, these agents could retrieve only information for a subset of agents depending on the event detected in a concrete instant. Furthermore, for the study of the motion vectors it could be considered the creation of a set of overlapping regions instead of using a strict partitioning of the scene.

References

1. Bellifemine, F., Caire. G., Greenwood, D.: Developing multi-agent systems with JADE. Wiley (2015)
2. Castan, J.A., Ibarra, S., Laria, J., Guzman, J., Castan, E.: Control de trafico basado en agentes inteligentes. Polibits **50**, 61–68 (2014)
3. Dresner, K., Stone, P.: A multi-agent approach to autonomous intersection management. J. Artif. Intell. Res. **31**, 591–656 (2008)
4. FIPA: Agent communication language specifications. Retrieved from http://www.fipa.org/repository/aclspecs.html, 02 June 2017
5. Huang, Y., Schmitt, F.G.: Time dependent intrinsic correlation analysis of temperature and dissolved oxygen time series using empirical mode decomposition. J. Mar. Syst. **130**, 90–100 (2014). ISSN 0924-7963

6. Loarte, R., Quizhpea, B., Paz-Arias, H.: Desarrollo y simulacion de un sistema multi-agente para la comunicacion de semaforos para encontrar la ruta optima mediante grafos. Rev. Tecnol. ESPOL **1**, 43–63 (2015)
7. Lorentzen, T.: Statistical analysis of temperature data sampled at Station-M in the Norwegian Sea. J. Mar. Syst. **130**, 31–45 (2014). ISSN 0924-7963
8. Moghtaderi, A., Flandrin, P., Borgnat, P.: Trend filtering via empirical mode decompositions. Comput. Stat. Data Anal. **58**, 114–126 (2013). ISSN 0167-9473
9. Mueen, A., Hamooni, H., Estrada, T.: Time series join on subsequence correlation. In: Proceedings of the 2014 IEEE International Conference on Data Mining (ICDM 2014), pp. 450–459. IEEE Computer Society, Washington, DC, USA (2014)
10. Olagnon, M., Agbko Kpogo-Nuwoklo, K., Gud, Z.: Statistical processing of West Africa wave directional spectra time-series into a climatology of swell events. J. Mar. Syst. **130**, 101–108 (2014). ISSN 0924-7963
11. Searle, J.R.: Speech Acts. Cambridge University Press (1969)

On the Antecedent Sets for Fuzzy Classification of Colorectal Polyps with Stabilized KH Interpolation

Szilvia Nagy, Ferenc Lilik and Laszlo T. Koczy

Abstract Polyps in the colorectal part of the bowel appear often, and in many cases these polyps can develop into malign lesions, such as cancer. Colonoscopy is the most efficient way to study the inner surface of the colorectum, and doctors usually are able to detect polyps on a motion picture diagnostic session. However, it is useful to have an automated tool that can help drawing attention to given parts of the image, and later a method for classification the polyps can also be developed. Statistical properties of the colour channels of the images are used as antecedents for a fuzzy decision system, together with edge densities and Renyi entropies-based structural entropy. However promising the processed images are, the variation in the preparation of the diagnosis as well as the practice of the operating personnel can lead to images with significantly different noise and distortion level, thus detecting the polyp can be complicated. In the following considerations image groups are presented that have similarities from the polyp detection point of view, and those type of images are also given, which can spoil a well prepared detecting system.

Keywords Fuzzy inference · Colorectal polyp · Fuzzy rule interpolation
Image segmentation

1 Introduction

Colorectal cancer is among those cancer types that are not easy to screen, as mostly endoscopy is needed for the diagnosis. It is also among the top 5 most lethal cancer types, mainly because of the insufficient screening and late diagnosis. This cancer

S. Nagy (✉) · F. Lilik · L. T. Koczy
Szechenyi Istvan University, Győr 9026, Hungary
e-mail: nagysz@sze.hu

F. Lilik
e-mail: lilikf@sze.hu

L. T. Koczy
Budapest University of Technology and Economics, Budapest 1117, Hungary
e-mail: koczy@tmit.bme.hu

© Springer Nature Switzerland AG 2019
M. E. Cornejo et al. (eds.), *Trends in Mathematics and Computational Intelligence*, Studies in Computational Intelligence 796,
https://doi.org/10.1007/978-3-030-00485-9_3

type is usually originated from polyps on the inner surface of the last part of the bowel, that are mostly visible with colonoscope, however, only a small percentage of the polyps has potential to develop into a malign object. The colorectal polyps are classified according to their shape, i.e., whether they are protruding into the bowel, slightly elevated, flat or depressed into the wall of the bowel [1, 2]. The pattern of the surface of the polyps can also be classified into groups from benign through the ones having potential to develop into malign, and pe-cancerous polyps to the developed cancer, each group having their characteristic surface-pit patterns [3].

Endoscopy is a method that is not likely to be used for large-scale screening of the population, as it is costly, the endoscope has to be operated by expensive medical professionals, and it is considered as inconvenient by most of the population. Capsular endoscopy seems to be more reasonable, however, it is still costly, the images are of poorer quality, and sometimes problems arise when passing through the bowel [4]. The polyps usually can be recognized by the operator of the endoscope, however, an automatic detection aid can provide help for the medical staff. Also automatic pattern recognition could lead to less expensive diagnosis, and less need for biopsy. Based on our experience with classification of telecommunication lines [5, 6], a fuzzy inference system was introduced with 15 antecedent dimensions for colonoscopic image segments. The antecedent dimensions were statistical parameters of the processed image [7, 8], that seemed to be relevant for distinguishing the image parts with polyp segments from those without polyp, however, as mostly the classification scheme often tended to classify even the segments with polyp into non-polyp class we studied the reason for this performance.

In the following considerations a fuzzy classification scheme is given for determining whether a given image segment contains polyp parts. As a first step, in Sect. 2 antecedent dimensions are selected that have the potential to distinguish between parts with polyp and without polyp. As a second step, in Sect. 3, Mamdani-type fuzzy classification method is outlined with stabilized KH interpolation. Next, based on the fuzzy membership functions of the antecedent dimensions, the classification method is given. In Sect. 4, the classification algorithm is applied to a database of images available at [9]. As the database consists of very different image types, the images are grouped, so that we could determine, which image types tend to "hide" their polyps, and for which images are the selected antecedent dimensions good. In the last section the results are discussed.

2   On the Selected Antecedent Dimensions

In colonoscopy images usually there are many unfavourable features. The images are mostly pink, with shiny surfaces and lots of reflections of the lightsource of the device. Being blurry due to movement of the bowel, covered with remaining material due to insufficient preparation of the patient are not uncommon, moreover, usually the camera resolution is not too high and the image compression method is optimized for motion picture view and medical report prints, thus making artifacts, tiling, colour distortion is also possible.

Due to these facts, as a first step we decided to select such parameters for the decision, that can distinguish a background with larger, bluish pattern (the veins) from the polyps, that are mostly lighter, have more dense, and not blue pattern, have a smaller curvature compared to the large bowel wall and has shadows and contour. As the polyps are mostly better lighted than its environment, simple statistical parameters, as the mean and the standard deviation can theoretically useful. Also, the edge density is larger in polyps, if the focus of the camera is not too bad, thus we selected a standard, Cranny filter based edge detection algorithm to convert the picture into black in general and white in placed where edges are present, and calculated the number of white pixels compared to the total number of pixels. As a last step, as we wanted information on the shape of the surface, we also selected structural entropy of the image segment.

2.1 On Structural Entropies

Pipek and Varga introduced [10–14] a pair of entropy based quantities (q, S_{str}) for describing the topology free structure of a function, that is normalized as a probability distribution, i.e. in our case the brightness of the pixels Q_n has to be normalized as

$$Q_n \geq 0, \quad \text{for } n = 1, \ldots, N \tag{1}$$

$$\sum_{n=1}^{N} Q_n = 1. \tag{2}$$

As a first step, a so called participation ratio or delocalization measure

$$D = \left(\sum_{n=1}^{N} Q_n^2 \right)^{-1} \tag{3}$$

was determined. This quantity shows, to how many pixels the distribution Q_n expands. The participation ratio can be normalized by the total number of pixels N, thus receiving another quantity, the so called spatial filling factor

$$q = \frac{D}{N}. \tag{4}$$

This gives the ratio of the pixels and fulfills the following inequality

$$\frac{1}{N} \leq q \leq 1. \tag{5}$$

Mostly the logarithm of the filling factor is used.

The well-known Shannon or von Neumann entropy

$$S = -\sum_{n=1}^{N} Q_n \ln Q_n \tag{6}$$

is a quantity that measures how much the pixel distribution $\{Q_n | n = 1, \ldots, N\}$ deviates from the uniform distribution (where all Q_ns are the same, $1/N$).

The Shannon entropy can be divided into two parts, the extension and the structural entropy. As the distribution $\{Q_n\}$ is localized to D lattice sites, the extension entropy can be defined as

$$S_{ext} = \ln D. \tag{7}$$

This quantity is practically the entropy of a uniform distribution over D sites. The remaining part, the structural entropy thus reflects on the shape of the distribution: it gives how much the distribution $\{Q_n\}$ differs from the step function. The definition of the structural entropy is

$$S_{str} = S - \ln D. \tag{8}$$

the inequality

$$0 \le S_{str} \le -\ln q \tag{9}$$

is valid for the structural entropy.

The pair of quantities (q, S_{str}) is called generalized localization. If a distribution $\{Q_n\}$ is of a given localization type (e.g., Gaussian, exponential), and its structural entropy is plotted as a function of its filling factor q, it is located on a characteristic line on the localization map $S_{str}(lnq)$. As an example, the localization map of the blue colour channel of the 18 groups of images from database [9] are given in Fig. 1.

3 On Fuzzy Classification with Interpolation of the Sparse Rule Bases

Since the time Zadeh introduced fuzzy sets [15] their popularity increased rapidly, as the concept of not belonging to a group entirely was very effective in decision and control systems and it can be applied with good results in pattern recognition, too [16]. Mamdani with his coworkers was one of the first persons, who developed fuzzy set based inference systems [17], which was effective in our previous work [6]. However, if the resulting rule bases are sparse, and the value of an observable is located outside of the supports of the rules, an interpolation has to be carried out in order to make the evaluation, and thus the inference possible.

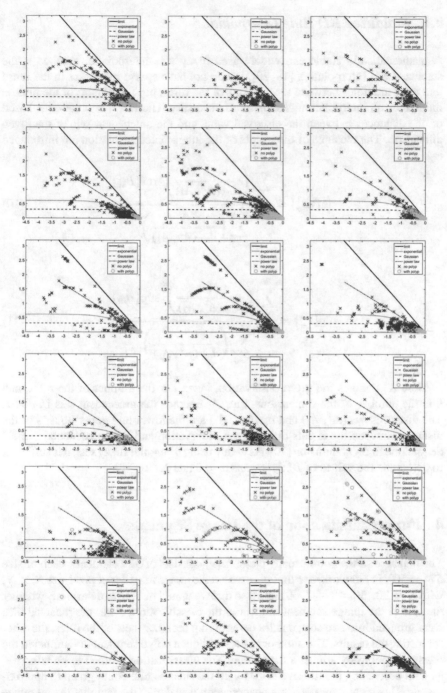

Fig. 1 Structural entropy plots S_{str} versus $\ln q$ for the 18 image groups for 50 by 50 tile size for the colour channel blue

3.1 Stabilized KH Rule Interpolation

A mathematically stable and widely used fuzzy rule interpolation method is the stabilized KH interpolation [18–20]. We do not have sparse rule bases in the sense that there are holes between the supports of the rules, but the support of the rules is finite, thus an interpolation is sometimes necessary. The KH interpolation is based on the distances between the observed value and the antecedent sets of the given dimension. The closures of the α-cuts of the interpolated resolution are introduced as

$$\inf\{B_\alpha^*\} = \frac{\sum_{i=1}^{2n}\left(\frac{1}{d_{\alpha L}(A^*, A_i)}\right)^k \inf\{B_{i\alpha}\}}{\sum_{i=1}^{2n}\left(\frac{1}{d_{\alpha L}(A^*, A_i)}\right)^k} \tag{10}$$

and

$$\sup\{B_\alpha^*\} = \frac{\sum_{i=1}^{2n}\left(\frac{1}{d_{\alpha U}(A^*, A_i)}\right)^k \sup\{B_{i\alpha}\}}{\sum_{i=1}^{2n}\left(\frac{1}{d_{\alpha U}(A^*, A_i)}\right)^k}. \tag{11}$$

Here A^* denotes the given observation, index i is the number of the rules, and k is the number of the dimensions, thus A_i denotes the antecedent sets in rule i. The distances $d_{\alpha L}(A^*, A_i)$ and $d_{\alpha U}(A^*, A_i)$ are the lower and upper bounds of the distance between the α-cuts of the given observation and the antecedents, and B^* denotes the fuzzy conclusion [20]. As we are using triangular fuzzy sets, it is sufficient to determine the values of B_α^* for α-cuts $\alpha = 0$ and 1.

4 Fuzzy Classification of the Image Segments

The database of the colonoscopy pictures [9] consists of 379 elements, each of size 574×500. In our work, as a first step, we cut the image into tiles of pixel size $N \times N$, with $N = 20, 30, 40, 50, 60$ and 70 so that we would be able to determine, whether the size of the image segment influences the classification. Next, the received tiles were grouped into two sets, one for determining the fuzzy sets of the rules, the other for testing the results. The number of tiles without a polyp segment outnumbered the ones with polyp by almost a magnitude. The classification seemed to be effective, as with decreasing the tile size to 20 by 20 the success rate could go up to approximately 87 percent, however, this phenomenon was due to the fact that the inference tended to classify everything to the non-polyp category, and the non-polyp tiles were approximately 5–9 times more than the ones with polyp (of course, 5 belonged to

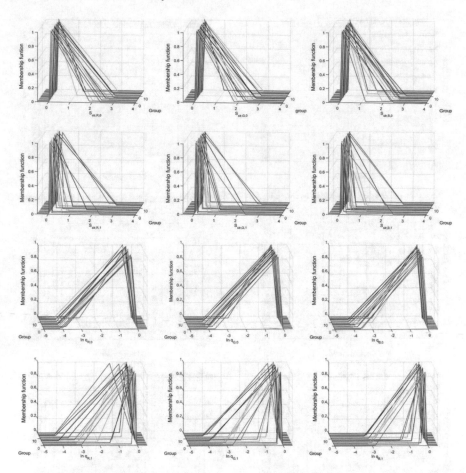

Fig. 2 Membership functions for the 6 entropy based antecedent dimensions, for the 18 image groups, in the case of tile size 50 by 50 pixels

the 70×70 tile size, and 9 to the 20×20). Also the three colour channels were separated, as the pit patterns were more visible in the blue and green channels, as well as the vein patterns were more emphasized in the green channels.

We investigated the reason for this bad behaviour. First, we studied the appearance of the images. There were images with clean, nice, clearly distinguishable bowel background, and sharp polyp edges, almost in focus, in some cases excess liquid of yellowish colour, or whitish plaque was present, in one set the colon was not cleaned beforehand, and in some cases the image was though of a cleaned bowel with ideal condition, the camera became out of focus throughout the whole process. There were multiply included pictures as well. In order to determine for what type of images our method can be used, we grouped the pictures into 18 groups, each belonging to the same polyp with similar conditions, just the distance and the angle were different.

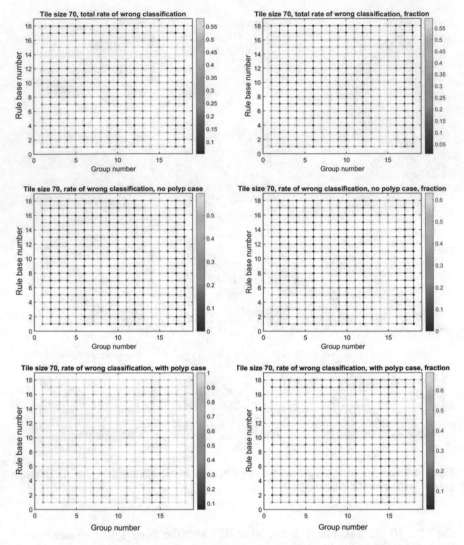

Fig. 3 Misclassification rate of the rule bases for 70 by 70 tile size for the 18 groups of images. Upper row shows the total misclassification rate, second row the cases without polyp, whereas lower row gives the misclassification rate in case of image sections wit polyp. The second columns shows the difference of the group the image was classified into and the percentage of the area of the image section that is covered with polyp

As it can be seen in Fig. 1, the groups had in some cases similar, some cases very different properties. We determined the membership functions for both the images with polyp and for those without by determining the minimum, maximum and mean values of the given parameter of the given group. The results for a medium, 50 by 50 tile size for the entropy dimensions are given in Fig. 2. In the figure the columns mean the colour channel, the first two rows are the membership functions of the structural entropy parameter for the case without polyp and with polyp. The second two rows are the logarithms of the filling factor for no-polyp and polyp cases. It can be seen that there are image groups where the membership functions of the tiles with polyp are almost the same as the membership functions of the ones without polyp, thus in these cases the structural entropy and filling factor are not good candidates for searching for polyps. There are cases, when the two antecedent sets are significantly different, but even in these cases the α-cuts where $\alpha = 1$ are almost in the same place, thus classifying a tile into a "non-polyp" class is highly probable based on this dimension. Using subnormal fuzzy set for the zero dimension might solve this problem, but selecting the value of the maximal non-empty α-cut is also a hard task.

Using all the mentioned antecedent dimensions, the mean, the standard deviation, the edge density and the localization parameters, we determined the rulebases for all the 18 picture groups. Using all these rulebases the classification for all the tiles in the test set was performed. The results are grouped according to the previously selected image groups and plotted in the first column of Fig. 3. The uppermost plot belongs to the general results, the second row belongs to only those image segments, where no polyps were present whereas the las row belongs to the images with polyp. It can be seen, that the images with polyp are still highly misclassified, especially in some cases (the most notable one is the image series with not sufficient preparation of the patient). The cases, where either the vein structure of the bowel wall, or the pit pattern of the polyp is clearer, the classification goes the best.

As the area of the tiles is not fully covered by the polyps, the polyp covered percentage is also calculated. The difference of the polyp coverage and the classification output is presented also in Fig. 3. It can be seen, that the success rate increases significantly, especially in some cases it goes beyond 80 percent even in the case of the pictures with polyp, which means, that the wrongly classified images contained mainly only in a small area polyps.

5 Conclusions

The first steps in selecting the antecedent dimensions of a novel colorectal polyp detecting method was presented. All the dimensions are calculated on tiles of the image of various sizes. The statistical parameters, such as mean, standard deviation and edge density are significantly different either in the case of clear, well prepared images, or in the case of overly lighted polyps. In the case of the clearly visible structures, structural entropies are also good candidates for antecedent dimensions. In the case of dirty or off-focus images, clearly other antecedent dimensions should be selected. We have also determined that from the several hundred thousand image segments, mainly those were falsely classified that contained polyp only in a small part of their area.

Acknowledgements The authors would like to thank the financial support of the project EFOP-3.6.2-16-2017-00015 HU MATHS—IN—Intensification of the activity of the Hungarian Industrial Innovation Mathematical Service Network, and the ÚNKP-17-4-III-SZE-16 New National Excellence Programme of the Ministry of Human Capacities of Hungary.

References

1. Søreide, K., Nedrebø, B.S., Reite, A., et al.: Endoscopy morphology, morphometry and molecular markers: predicting cancer risk in colorectal adenoma. Expert Rev. Mol. Diagn **9**, 125–137 (2009)
2. Jass, J.R.: Classification of Colorectal Cancer Based on Correlation of Clinical, Morphological and Molecular Features, Histopathology, vol. 50, pp. 113–130. Wiley (2006)
3. Kudo, S., Hirota, S., Nakajima, T.: Colorectal tumours and pit pattern. J. Clin. Pathol. **47**, 880–885 (1994)
4. Rácz, I., Jánoki, M., Saleh, H.: Colon cancer detection by Rendezvous Colonoscopy: successful removal of stuck colon capsule by conventional colonoscopy. Case Rep. Gastroenterol. **4**, 19–24 (2010)
5. Lilik, F., Kóczy, L.T.: The determination of the Bitrate on Twisted Pairs by Mamdani inference method. In: Issues and Challenges of Intelligent System and Computational Intelligence, Studies in Computational Intelligence, vol. 530, pp. 59–74 (2014). https://doi.org/10.1007/978-3-319-03206-1_5
6. Lilik, F., Nagy, S., Kóczy, L.T.: Wavelet based fuzzy rule bases in pre-qualification of access networks wire Pairs. In: IEEE Africon 2015, Addis Ababa, Ethiopia, 14–17 September 2015
7. Georgieva, V.M., Nagy, S., Kamenova, E., Horváth, A.: An approach for pit pattern recognition in colonoscopy images. Egypt. Comput. Sci. J. **39**, 72–82 (2015)
8. Georgieva, V.M., Vassilev, S.G.: Kidney Segmantation in ultrasound images via active contours. In: 11th International Conference on Communications, Electromagnetics and Medical Applications, Athens, Greece, October 2016
9. Bernal, J., Sanchez, F.J., Vilariño, F.: Towards automatic polyp detection with a polyp appearance model. Pattern Recognit. **45**, 3166–3182 (2012)
10. Pipek, J., Varga, I.: Universal classification scheme for the spatial localization properties of one-particle states in finite d-dimensional systems. Phys. Rev. A **46**, 3148–3164 (1992). APS, Ridge NY-Washington DC
11. Varga, I., Pipek, J.: Ranyi entropies characterizing the shape and the extension of the phase space representation of quantum wave functions in disordered systems. Phys. Rev. E **68**, 026202 (2003). APS, Ridge NY-Washington DC
12. Molnár, L.M., Nagy, S., Mojzes, I.: Structural entropy in detecting background patterns of AFM images. Vacuum **84**, 179–183 (2010). Elsevier, Amsterdam
13. Bonyár, A., Molnár, L.M., Harsányi, G.: Localization factor: a new parameter for the quantitative characterization of surface structure with atomic force microscopy (AFM). Micron **43**, 305–310 (2012). Elsevier, Amsterdam
14. Bonyár, A.: AFM characterization of the shape of surface structures with localization factor. Micron **87**, 1–9 (2016)
15. Zadeh, L.A.: Fuzzy sets. Inf. Control **8**, 338–353 (1965). https://doi.org/10.1016/S0019-9958(65)90241-X
16. Tormási, A.: Fuzzy-based, adaptive, online single-stroke recognizer. In: The Third Gyr Symposium on Computational Intelligence: Book of Abstracts. Győr, Hungary, 28–29 September, pp. 19–20 (2010)
17. Mamdani, E.H., Assilian, S.: An experiment in linguistic synthesis with a fuzzy logic controller. Int. J. Man-Mach. Stud. **7**, 113 (1975). https://doi.org/10.1016/S0020-7373(75)80002-2

18. Kóczy, L.T., Hirota, K.: Approximate reasoning by linear rule interpolation and general approximation. Int. J. Approx. Reason. **9**, 197–225 (1993). https://doi.org/10.1016/0888-613X(93)90010-B
19. Kóczy, L.T., Hirota, K.: Interpolative reasoning with insufficient evidence in sparse fuzzy rule bases. Inf. Sci. **71**, 169–201 (1993). https://doi.org/10.1016/0020-0255(93)90070-3
20. Tikk, D., Joó, I., Kóczy, L.T., Várlaki, P., Moser, B., Gedeon, T.D.: Stability of interpolative fuzzy KH-controllers. Fuzzy Sets Syst. **125**, 105–119 (2002). https://doi.org/10.1016/S0165-0114(00)00104-4

How to Predict Consistently?

Evgeni Vityaev and Sergei Odintsov

Abstract One of reasons for arising the statistical ambiguity is using in the course of reasoning laws which have probabilistic, but not logical justification. Carl Hempel supposed that one can avoid the statistical ambiguity if we will use in the probabilistic reasoning maximal specific probabilistic laws. In the present work we deal with laws of the form $\varphi \Rightarrow \psi$, where φ and ψ are arbitrary propositional formulas. Given a probability on the set of formulas we define the notion of a maximal specific probabilistic law. Further, we define a prediction operator as an inference with the help of maximal specific laws and prove that applying the prediction operator to some consistent set of formulas we obtain a consistent set of consequences.

Keywords Probabilistic inference · Maximal specificity · Prediction
Consistency

1 Introduction

The statistical ambiguity problem arises due to using in the course of reasoning laws which have a probabilistic, but not logical justification. Carl Hempel supposed that one can avoid the statistical ambiguity if we will use in the probabilistic reasoning only so called maximal specific probabilistic laws (see Sect. 2). In the present work we deal with laws of the form $\varphi \Rightarrow \psi$, where φ and ψ are arbitrary propositional formulas. In Sect. 3 we define the concept of probability on the set of formulas close to that of [1, 8] and extend it to the family of rules. Finally, in Sect. 4 we define the set of maximal specific probabilistic laws and the prediction operator as an inference with the help of maximal specific laws. Then we prove that applying the prediction

E. Vityaev · S. Odintsov (✉)
Sobolev Institute of Mathematics, Novosibirsk, Russian Federation
e-mail: odintsov@math.nsc.ru

E. Vityaev
e-mail: vityaev@math.nsc.ru

© Springer Nature Switzerland AG 2019
M. E. Cornejo et al. (eds.), *Trends in Mathematics and Computational
Intelligence*, Studies in Computational Intelligence 796,
https://doi.org/10.1007/978-3-030-00485-9_4

operator to a consistent set of formulas we obtain the set of consequences, which is consistent too.

Despite the explanation and the prediction have the same logical structure, we prefer the term "prediction" in the following cases: if the act of prediction precedes a predicted fact in time. We also speak on a prediction if a predicted fact remains unknown due to different reasons: too high costs of establishing this fact, an impossibility to establish the fact at the moment, the lack of this fact in a series of preceding experiments, etc.

2 Statistical Ambiguity and Requirement of Maximal Specificity

The *Covering Law Model* suggested by Carl Hempel [3] (see [4, 7] for a historical overview) distinguished two kinds of explanation: Deductive-Nomological explanations (D-N explanations) and Inductive-Statistical explanations (I-S explanations). A D-N argument is a standard logical inference of facts from other facts with the help of general laws. An I-S argument has the form:

$$\frac{\begin{matrix} p(G; F) = t \\ F(a) \end{matrix}}{G(a)}$$

The line distinguishes the *explanandum* $G(a)$ from two premises (*explanans*), one of which has the form of a statistical law of the form $p(G; F) = t, 0 \le t \le 1$, where t denotes the probability that an object from the set defined by predicate F is also a member of the set defined by predicate G.

Right from the beginning it was clear to Hempel that two I-S explanations can yield contradictory conclusions. He called this phenomenon the statistical ambiguity of I-S explanations [4]. Recall one of traditional examples of the statistical ambiguity. Suppose that we have the following statements.

L1 *Almost all cases of streptococcus infection clear up quickly after the administration of penicillin.*

L2 *Almost no cases of penicillin resistant streptococcus infection clear up quickly after the administration of penicillin.*

C1 *Jane Jones had streptococcus infection.*

C2 *Jane Jones received treatment with penicillin.*

C3 *Jane Jones had a penicillin resistant streptococcus infection.*

Following the above pattern one can construct two I-S explanations based on these statements. On the base of L1 and C1∧C2 one can explain why Jane Jones recovered quickly (E). The second argument with premises L2 and C2∧C3 explains why Jane Jones did not (¬E). The set of premises {C1, C2, C3} is consistent. However, the conclusions contradict each other, making these arguments rival ones.

Hempel hoped to solve this problem by forcing all statistical laws in an argument to be maximally specific—they should contain all relevant information with respect to the domain in question. In our example, then, the premiss $C3$ invalidates the first argument, since this argument is not maximally specific with respect to all information about Jane Jones. So, we can only explain ¬E, but not E.

In [4] Hempel defined the Requirement of Maximal Specificity (RMS) as follows. An I-S argument

$$p(G; F) = t$$
$$\frac{F(a)}{G(a)}$$

is an acceptable I-S explanation with respect to a "knowledge state" K, if the following Requirement of Maximal Specificity is satisfied. For any predicate H for which the following two sentences are contained in K: $\forall x(H(x) \Rightarrow F(x))$, $H(a)$, there exists a statistical law $p(G; H) = t'$ in K such that $t = t'$. The basic idea of RMS is that if F and H both contain the object a, and H is a subset of F, then H provides more specific information about the object a than F, and therefore the law $p(G; H)$ should be preferred over the law $p(G; F)$. However the law $p(G; H)$ has the same probability as the law $p(G; F)$.

3 Probability on Propositional Formulas and Rules

Our attention will be restricted to propositional logic. We start from a set of atoms At and construct from them the set of well formed formulas $F(At)$ using connectives $\wedge, \vee, \rightarrow, \neg$. As usual the equivalence $\leftrightarrow$ is considered as an abbreviation. We define $\top$ as $\varphi \vee \neg\varphi$, where φ is some fixed formula. For a finite set of formulas T the conjunction of its elements is denoted by $\bigwedge T$. By $V(\varphi)$ we denote the set of atoms occuring in formula φ. The set $F(At)$ with naturally interpreted connectives forms an algebra of formulas $\mathcal{F}(At)$. The classical interpretation of propositional connectives is assumed, therefore models for our logic can be identified with mappings from At to the set of classical truth values $\{0, 1\}$. We call such mappings $(At\text{-})valuations$. Every valuation $v : At \rightarrow \{0, 1\}$ extends in a standard way to the set $F(At)$ using classical truth tables for connectives $\wedge, \vee, \rightarrow$, and $\neg$, the extended valuation we denote in the same way $v : F(At) \rightarrow \{0, 1\}$. A formula φ is $satisfiable\ in$ a set $\mathfrak{G}$ of valuations if $v(\varphi) = 1$ for some $v \in \mathfrak{G}$, a formula φ $holds\ on$ $\mathfrak{G}$, $\mathfrak{G} \models \varphi$, if $v(\varphi) = 1$ for all $v \in \mathfrak{G}$. Finally, a set $T \subseteq F(At)$ is $\mathfrak{G}\text{-}consistent$ if there is $v \in \mathfrak{G}$ such that $v(\varphi) = 1$ for all $\varphi \in T$. The set of all valuations we denote $\mathfrak{All}$.

Let $\mathfrak{G} \subseteq \mathfrak{All}$. The relation $\varphi \equiv_{\mathfrak{G}} \psi$ is defined by the condition that $\varphi \leftrightarrow \psi$ holds on $\mathfrak{G}$. It is clear that $\equiv_{\mathfrak{G}}$ is a congruence on $\mathcal{F}(At)$, the respective quotient is denoted as $\mathcal{B}^{\mathfrak{G}}(At)$. The coset of φ w.r.t. $\equiv_{\mathfrak{G}}$ is denoted as $[\varphi]_{\mathfrak{G}}$. Recall that the universe of $\mathcal{B}^{\mathfrak{G}}(At)$ equals $\{[\varphi]_{\mathfrak{G}} \mid \varphi \in F(At)\}$ and that this set is finite whenever At is finite. The operations of $\mathcal{B}^{\mathfrak{G}}(At)$ are denoted as $\wedge_{\mathfrak{G}}, \vee_{\mathfrak{G}}, \rightarrow_{\mathfrak{G}}, \neg_{\mathfrak{G}}$ and the lattice order as $\sqsubseteq_{\mathfrak{G}}$. Recall that $[\varphi]_{\mathfrak{G}} \sqsubseteq_{\mathfrak{G}} [\psi]_{\mathfrak{G}}$ iff $[\varphi]_{\mathfrak{G}} = [\varphi \wedge \psi]_{\mathfrak{G}} = [\varphi]_{\mathfrak{G}} \wedge_{\mathfrak{G}} [\psi]_{\mathfrak{G}}$.

Let $\mu : 2^{\mathfrak{G}} \to [0, 1]$ be a finitely additive measure defined on $\mathfrak{G}$, i.e., μ is such that: (1) $\mu(\mathfrak{G}) = 1$; 2) $\mu(\varnothing) = 0$; and (3) $\mu(A_1 \cup \ldots \cup A_n) = \mu(A_1) + \ldots + \mu(A_n)$ for pairwise disjoint subsets $A_1, \ldots, A_n \subseteq \mathfrak{G}$. Additionally we assume that $\mu(A) = 0$ implies $A = \varnothing$. Elements of $\mathfrak{G}$ may be interpreted as outcomes of experiments. Further, we assume that only essential experiments are included in $\mathfrak{G}$, which explains why $\mu(\{v\}) \neq 0$ for all $v \in \mathfrak{G}$.

For every $\varphi \in F(At)$, put $\varphi^{\mathfrak{G}} := \{v \in \mathfrak{G} \mid v(\varphi) = 1\}$. Now we define a function $\nu : F(At) \to [0, 1]$ by the rule $\nu(\varphi) = \mu(\varphi^{\mathfrak{G}})$. It is easy to check that ν satisfies the following properties.

1. $\nu(\varphi) = 1$ iff φ is holds on $\mathfrak{G}$.
2. $\nu(\varphi) = 0$ iff φ is not satisfiable on $\mathfrak{G}$.
3. $\nu(\varphi \vee \psi) = \nu(\varphi) + \nu(\psi)$ iff $\varphi \wedge \psi$ is not satisfiable in $\mathfrak{G}$.

This means that ν is a *probability* on the set of propositional formulas in a sense close to that of [1, 8].

Now we generalize the notions from [9]. By a *rule* we mean a syntactic object of the form

$$r = \varphi \Rightarrow \psi,$$

where $\varphi, \psi \in F(At)$ and $\varphi \to \psi$ is not a logical tautology. We call φ and ψ a *body* and a *head* of r: $\varphi = B(r)$ and $\psi = H(r)$. The probability of a rule $r = \varphi \Rightarrow \psi$ with $\nu(\varphi) \neq 0$ is defined as follows: $\nu(r) := \nu(\psi|\varphi) = \frac{\nu(\psi \wedge \varphi)}{\nu(\varphi)}$. In case, φ is not satisfiable on $\mathfrak{G}$, the value $\nu(r)$ remains undefined. Notice that the value $\nu(r)$ was defined so that it is smaller or equal to $\nu(\varphi \to \psi)$.

Definition 1 Let r_1 and r_2 be two rules with the same head, $H(r_1) = H(r_2)$. We call r_2 a *generalization* of r_1, symbolically $r_2 \succeq r_1$ if $B(r_1)^{\mathfrak{G}} \subseteq B(r_2)^{\mathfrak{G}}$; rule r_2 is a *proper* generalization of r_1, $r_2 \succ r_1$, if $r_2 \succeq r_1$ and $B(r_1)^{\mathfrak{G}} \neq B(r_2)^{\mathfrak{G}}$. We say in this case that r_1 is a *(proper) specialization* of r_2.

In other words, one of the two rules with the same head is a proper generalization of the other if its body is weaker from the logical point of view.

4   Prediction Operator

In this section we generalize the results of [10].[1] Let At, a set $\mathfrak{G}$ of valuations, and a measure μ on G be fixed. Assume that $\mathcal{R}$ is a set of rules such that for $r \in \mathcal{R}$ and a rule s such that $s \prec r$ and $\nu(s) > \nu(r)$, there is $r' \in \mathcal{R}$ with $r' \preceq s$. Now we introduce two special subsets of $\mathcal{R}$:

$$\mathsf{M}_1(\mathcal{R}) = \{r \in \mathcal{R} \mid (\top \Rightarrow H(r)) \succ r \text{ implies } \nu(r) > \nu(\top \Rightarrow H(r))\};$$

---

[1]In [10], rules are of the form $\alpha_1 \wedge \ldots \wedge \alpha_n \Rightarrow \beta$, where $\alpha_1, \ldots, \alpha_n, \beta$ are literals, i.e., atoms or negations of atoms.

$$M_2(\mathcal{R}) = \{r \in \mathcal{R} \mid r \in M_1(\mathcal{R}) \text{ and } \forall r' \in M_1(\mathcal{R})(r \succ r' \Rightarrow \nu(r') \leq \nu(r))\}.$$

Namely rules from $M_2(\mathcal{R})$ we consider as satisfying the Requirement of Maximal Specificity, because a specification of such rule by a rule from $M_1(\mathcal{R})$ does not lead to an increase of probability, which means that for $r \in M_2(\mathcal{R})$, its body contains all statistically relevant information for the prediction of $H(r)$.

For a set of rules $\Pi \subseteq M_2(\mathcal{R})$ we define an operator of *direct predictions*:

$$Pr_\Pi(T) = T \cup \{H(r) \mid r \in \Pi, \exists \varphi_1, ... \varphi_n \in T (\mathfrak{G} \models (\varphi_1 \wedge ... \wedge \varphi_n) \leftrightarrow B(r))\},$$

where T is a set of formulas. Further, we put:

$$Pr_\Pi^0(T) = T, \ Pr_\Pi^{n+1}(T) = Pr_\Pi(Pr_\Pi^n(T)), \ PR_\Pi(T) = \bigcup_{n \in \omega} Pr_\Pi^n(T).$$

It is clear that $PR_\Pi(T)$ is the least fixed point of the operator of direct predictions containing T. We call PR_Π a *prediction operator for* Π.

Theorem 1 *Let At be a finite set of atoms, $\Pi \subseteq M_2(\mathcal{R})$, and $T \subseteq F(At)$. If T is a $\mathfrak{G}$-consistent set of formulas, then $PR_\Pi(T)$ is $\mathfrak{G}$-consistent.*

Proof Obviously, it will be enough to check that the operator of direct predictions produces a $\mathfrak{G}$-consistent set of formulas. Further, since At is finite, the family of all formulas is finite up to equivalence, and we may assume that sets T and Π are finite too.

Let $\Pi' = \{r_0, \ldots, r_n\}$ be the set of such rules r from Π that the equivalence $(\varphi_1 \wedge ... \wedge \varphi_n) \leftrightarrow B(r)$ holds on $\mathfrak{G}$ for some $\varphi_1, ..., \varphi_n \in T$. We put $T_0 = T, T_{i+1} = T_i \cup \{H(r_i)\}$. Clearly, $T_n = Pr_\Pi(T)$. Now we prove by induction that every T_i is $\mathfrak{G}$-consistent.

Assume that T_i is $\mathfrak{G}$-consistent, but T_{i+1} is not. Let $r_i = \varphi \Rightarrow \psi$. By definition of Π' there are $\chi_1, \ldots, \chi_n \in T$ such that $(\chi_1 \wedge ... \wedge \chi_n) \leftrightarrow \varphi$ holds on $\mathfrak{G}$. Let $N = T_i \setminus \{\chi_1, ... \chi_n\}$. Assume that $\{\varphi, \neg(\bigwedge N)\}$ is $\mathfrak{G}$-consistent, i.e., $\nu(\varphi \wedge \neg(\bigwedge N)) \neq 0$. In this case for $s = \varphi \wedge \neg(\bigwedge N) \Rightarrow \psi$ we have:

$$\nu(s) = \frac{\nu(\varphi \wedge \neg(\bigwedge N) \wedge \psi)}{\nu(\varphi \wedge \neg(\bigwedge N))} = \frac{\nu(\varphi \wedge \psi) - \nu(\varphi \wedge \bigwedge N \wedge \psi)}{\nu(\varphi) - \nu(\varphi \wedge \bigwedge N)}.$$

We have $\mathfrak{G} \models \bigwedge T_{i+1} \leftrightarrow (\varphi \wedge \bigwedge N \wedge \psi)$ and $\mathfrak{G} \models \bigwedge T_i \leftrightarrow (\varphi \wedge \bigwedge N)$ by choice of $\chi_1, \ldots, \chi_n$. Since by assumption $\nu(\bigwedge T_{i+1}) = 0$ and $\nu(\bigwedge T_i) \neq 0$, we conclude that $\nu(\varphi \wedge \bigwedge N \wedge \psi) = 0$ and $\nu(\varphi \wedge \bigwedge N) \neq 0$. In this way, we have

$$\nu(s) = \frac{\nu(\varphi \wedge \psi)}{\nu(\varphi) - \nu(\varphi \wedge \bigwedge N)} > \frac{\nu(\varphi \wedge \psi)}{\nu(\varphi)} = \nu(r_i).$$

Since $v(s) > v(r_i)$ there is $r' \in \mathcal{R}$ such that $r' \preceq s$. It follows from $v(r') > v(r_i)$ and $r_i \in M_1(\mathcal{R})$ that $r' \in M_1(\mathcal{R})$. On the other hand, from $r_i \in M_2(\mathcal{R})$ and $r_i \succ r'$ we obtain $v(r_i) \geq v(r')$. This contradiction proves that the body of s is not $\mathfrak{G}$-consistent: $v(\varphi \wedge \neg(\bigwedge N)) = 0$. As a consequence we obtain $v(\varphi \wedge \neg(\bigwedge N) \wedge \psi) = 0$. Now we have:

$$v(\varphi \wedge \psi) = v(\varphi \wedge \psi) - v(\varphi \wedge \neg(\bigwedge N) \wedge \psi) = v(\varphi \wedge \bigwedge N \wedge \psi) = 0.$$

Thus, $v(r_i) = 0$. At the same time $r_i \in M_1(\mathcal{R})$, which implies $0 = v(r_i) > v(\top \Rightarrow \psi) \geq 0$. The obtained contradiction concludes the proof.

Putting $\mathcal{R}$ to be the family of all rules trivializes the above statement, because in this case the prediction operator turns into a consequence operator over $\mathfrak{G}$. Refinement Theorem [10] shows that the class of rules considered in [10] satisfies the requirements imposed on $\mathcal{R}$. Other non-trivial cases of $\mathcal{R}$ will be considered in subsequent papers.

Of course, we worked in the ideal situation assuming that the probability on the set of formulas is known. In reality we have only statistical approximation of probabilities. The concept of semantical probabilistic inference (see [9]) aided at the search for maximal specific rules (of the form $\bigwedge_{i=1}^{n} \alpha_i \Rightarrow \beta$, where α_i and β are literals) on the base of statistically verified data gives a well-working approximation of $M_2(\mathcal{R})$-rules. This search procedure was realized in the program system Discovery [5]. The description of applications of this system to financial forecasting and to medicine can be found in [5, 6].

Acknowledgements The first of the authors (Sects. 1 and 2, also a coauthor of Theorem 1) was supported by the Russian Science Foundation (project # 17-11-01176). Both authors are grateful to the anonymous referees for their helpful reports and to participants of ESCIM'17 for the interesting discussion.

References

1. Fagin, R., Halpern, J.Y., Megiddo, N.: A logic for reasoning about probabilities. Inform. Comput. **80**, 78–128 (1990)
2. Fetzer, J.H.: Scientific Explanation. D. Reidel, Dordrecht (1981)
3. Fetzer, J.: Carl Hempel. In: Zalta, E.N. (ed.) Stanford Enciclopedia of Philosophy. Stanford University (2014). https://plato.stanford.edu/entries/hempel
4. Hempel, C.G.: Aspects of scientific explanation. In: Hempel, C.G. (ed.) Aspects of Scientific Explanation and other Essays in the Philosophy of Science. The Free Press, New York (1965)
5. Kovalerchuk, B., Vityaev, E.: Data Mining in Finance: Advances in Relational and Hybrid methods, 308 pp. Kluwer Academic Publishers (2000)
6. Kovalerchuk, B., Vityaev, E., Ruiz, J.F.: Consistent and complete data and "expert" mining in medicine. In: Medical Data Mining and Knowledge Discovery, pp. 238–280. Springer (2001)
7. Salmon, W.C.: Four Decades of Scientific Explanation. University of Minnesota Press, Minneapolis (1990)

8. Scott, D., Krauss P.: Assigning probabilities to logical formulas. In: Hintikka, J., Suppes, P. (eds.) Aspects of Inductive Logic, pp. 219–264. North-Holland (1966)
9. Vityaev, E.E.: The logic of prediction. In: Proceedings of the 9th Asian Logic Conference, Novosibirsk, Russia, 16–19 August 2006, pp. 263–276. World Scientific (2006)
10. Vityaev, E.E., Martynovich, V.V.: Probabilistic formal concepts with negation. In: Voronkov, A., Virbitskaite, I. (eds.) PSI 2014. LNCS, vol. 8974, pp. 1–15. Springer (2015)

Symbolic Unfolding of Multi-adjoint Logic Programs

Ginés Moreno, Jaime Penabad and José Antonio Riaza

Abstract The unfolding transformation has been widely used in many declarative frameworks for improving the efficiency of programs after applying computational steps on their rules. In this paper we apply such operation to a symbolic extension of a powerful fuzzy logic language where program rules extend the classical notion of clause by adding concrete and "symbolic" fuzzy connectives and truth degrees on their bodies.

Keywords Fuzzy logic programming · Symbolic programs · Unfolding

1 Introduction

During the last decades, several fuzzy logic programming systems have been developed. Here, essentially, the classical SLD resolution principle of logic programming has been replaced by a fuzzy variant with the aim of dealing with partial truth and reasoning with uncertainty in a natural way. Most of these systems implement (extended versions of) the resolution principle introduced by Lee [9], such as **Elf-Prolog** [3], **F-Prolog** [10] Fril [1], **MALP** [11], and **FASILL** [7].

In this paper we focus on the so-called *multi-adjoint logic programming* approach **MALP** [11], a powerful and promising approach in the area of fuzzy logic program-

This work has been partially supported by the EU (FEDER), the State Research Agency (AEI) and the Spanish *Ministerio de Economía y Competitividad* under grant TIN2016-76843-C4-2-R (AEI/FEDER, UE).

G. Moreno (✉) · J. A. Riaza
Department of Computing Systems, UCLM, 02071 Albacete, Spain
e-mail: Gines.Moreno@uclm.es

J. A. Riaza
e-mail: JoseAntonio.Riaza@uclm.es

J. Penabad
Department of Mathematics, UCLM, 02071 Albacete, Spain
e-mail: Jaime.Penabad@uclm.es

© Springer Nature Switzerland AG 2019
M. E. Cornejo et al. (eds.), *Trends in Mathematics and Computational Intelligence*, Studies in Computational Intelligence 796,
https://doi.org/10.1007/978-3-030-00485-9_5

ming. When specifying a MALP or any other fuzzy logic program managing fuzzy connectives and truth degrees beyond the simpler case of {*true*, *false*}, it might sometimes be difficult to assign weights—truth degrees—to program rules, as well as to determine the right connectives. This is a common problem with fuzzy control system design, where some trial-and-error is often necessary. In our context, a programmer can develop a prototype and repeatedly execute it until the set of answers is the intended one. Unfortunately, this is a tedious and time consuming operation. Actually, it might be impractical when the program should correctly model a large number of test cases provided by the user. In order to overcome this drawback, in [13–15] we have recently introduced a symbolic extension of MALP programs called *symbolic multi-adjoint logic programming* (sMALP).

Here, we can write rules containing *symbolic* truth degrees and *symbolic* connectives, i.e., connectives which are not defined on its associated multi-adjoint lattice. In order to evaluate these programs, we consider a symbolic operational semantics that delays the evaluation of symbolic expressions [13–15]. Therefore, a *symbolic answer* could now include symbolic (unknown) truth values and connectives. The approach is correct in the sense that using the symbolic semantics and then replacing the unknown values and connectives by concrete ones gives the same result as replacing these values and connectives in the original sMALP program and, then, applying the concrete semantics on the resulting MALP program. Furthermore, in [13–15] we show how sMALP programs can be used to tune a program w.r.t. a given set of test cases, thus easing what is considered the most difficult part of the process:

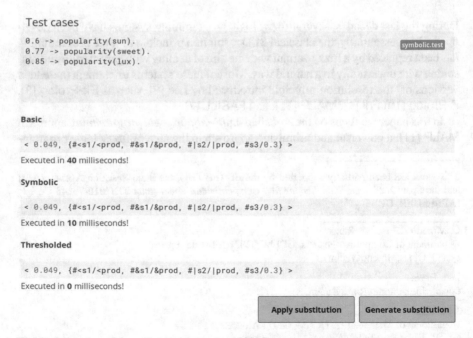

Fig. 1 Screenshot of the online tool for tuning sMALP programs

the specification of the right weights and connectives for each rule. Figure 1 mirrors the online implementation of this technique, which is freely available from http://dectau.uclm.es/tuning/.

The structure of this paper is as follows. After introducing in Sect. 2 the syntax of the framework of symbolic multi-adjoint logic programming, Sect. 3 focuses on the unfolding transformation we have specially tailored for this kind of fuzzy programs. We show that the formulation of this semantics preserving operation –which is initially devoted to produce more efficient code– strongly depends on the operational semantics formally defined for the framework. Finally, Sect. 4 concludes and points out some directions for further research.

2 Symbolic Multi-adjoint Logic Programs

We assume the existence of a multi-adjoint lattice $\langle L, \preceq, \&_1, \leftarrow_1, \ldots, \&_n, \leftarrow_n \rangle$, equipped with a collection of *adjoint pairs* $\langle \&_i, \leftarrow_i \rangle$—where each $\&_i$ is a conjunctor which is intended to be used for the evaluation of *modus ponens* [11]—. For instance and as shown in Fig. 2, we have typically several adjoint pairs over real numbers in the unit interval belonging to the well-known *Łukasiewicz's logic* $\langle \&_L, \leftarrow_L \rangle$, *Gödel's logic* $\langle \&_G, \leftarrow_G \rangle$ and *product's logic* $\langle \&_P, \leftarrow_P \rangle$, which might be used for modeling *pessimist*, *optimist* and *realistic scenarios*, respectively.

In addition, on each program rule, we can have a different adjoint implication ($\leftarrow_i$), conjunctions (denoted by $\wedge_1, \wedge_2, \ldots$), adjoint conjunctions ($\&_1, \&_2, \ldots$), disjunctions ($|_1, |_2, \ldots$), and other operators called aggregators (usually denoted by $@_1, @_2, \ldots$); see [16] for more details. More exactly, a multi-adjoint lattice fulfills the following properties:

- $\langle L, \preceq \rangle$ is a (bounded) complete lattice.
- For each truth function of $\&_i$, an increase in any of the arguments results in an increase of the result (they are *increasing*).
- For each truth function of $\leftarrow_i$, the result increases as the first argument increases, but it decreases as the second argument increases (they are *increasing* in the consequent and *decreasing* in the antecedent).
- $\langle \&_i, \leftarrow_i \rangle$ is an *adjoint pair* in $\langle L, \preceq \rangle$, namely, for any $x, y, z \in L$, we have that: $x \preceq (y \leftarrow_i z)$ if and only if $(x \&_i z) \preceq y$.

$\&_P(x,y) \triangleq x * y$	$\leftarrow_P (x,y) \triangleq \begin{cases} 1 & \text{if } y \leq x \\ x/y & \text{if } 0 < x < y \end{cases}$	*Product logic*
$\&_G(x,y) \triangleq \min(x,y)$	$\leftarrow_G (x,y) \triangleq \begin{cases} 1 & \text{if } y \leq x \\ x & \text{otherwise} \end{cases}$	*Gödel logic*
$\&_L(x,y) \triangleq \max(0, x+y-1)$	$\leftarrow_L (x,y) \triangleq \min(x-y+1, 1)$	*Łukasiewicz logic*

Fig. 2 Adjoint pairs of three standard fuzzy logics over $\langle [0,1], \leq \rangle$

In this work, given a multi-adjoint lattice L, we consider a first order language $\mathcal{L}_L$ built upon a signature Σ_L, that contains the elements of a countably infinite set of variables $\mathcal{V}$, function and predicate symbols (denoted by $\mathcal{F}$ and Π, respectively) with an associated arity—usually expressed as pairs f/n or p/n, respectively, where n represents its arity—, and the truth degree literals Σ_L^T and connectives Σ_L^C from L. Therefore, a well-formed formula in $\mathcal{L}_L$ can be either:

- A *value* $v \in \Sigma_L^T$, interpreted as itself, i.e., as the truth degree $v \in L$.
- $p(t_1, \ldots, t_n)$, if $t_1, \ldots, t_n$ are terms over $\mathcal{V} \cup \mathcal{F}$ and p/n is an n-ary predicate. This formula is called *atomic* (atom, for short).
- $\zeta(e_1, \ldots, e_n)$, if $e_1, \ldots, e_n$ are well-formed formulas and ζ is an *n*-ary connective with truth function $[\![\zeta]\!] : L^n \mapsto L$.

As usual, a *substitution* σ is a mapping from variables from $\mathcal{V}$ to terms over $\mathcal{V} \cup \mathcal{F}$ such that $Dom(\sigma) = \{x \in \mathcal{V} \mid x \neq \sigma(x)\}$ is its domain. Substitutions are usually denoted by sets of mappings like, e.g., $\{x_1/t_1, \ldots, x_n/t_n\}$. Substitutions are extended to morphisms from terms to terms in the natural way. The identity substitution is denoted by id. The composition of substitutions is denoted by juxtaposition, i.e., $\sigma\theta$ denotes a substitution δ such that $\delta(x) = \theta(\sigma(x))$ for all $x \in \mathcal{V}$.

A MALP *rule* over a multi-adjoint lattice L is a formula $H \leftarrow_i \mathcal{B}$, where H is an *atomic formula* (usually called the *head* of the rule), $\leftarrow_i$ is an implication symbol belonging to some adjoint pair of L, and $\mathcal{B}$ (which is called the *body* of the rule) is a well-formed formula over L without implications. A *goal* is a body submitted as a query to the system. A MALP program is a set of expressions R *with* v, where R is a rule and v is a *truth degree* (a value of L) expressing the confidence of a programmer in the truth of rule R. By abuse of the language, we often refer to R *with* v as a rule (see, e.g., [11] for a complete formulation of the MALP framework).

We are now ready for summarizing the *symbolic* extension of multi-adjoint logic programming initially presented in [13] where, in essence, we allow some undefined values (truth degrees) and connectives in the program rules, so that these elements can be systematically computed afterwards. In the following, we will use the abbreviation sMALP to refer to programs belonging to this setting.

Here, given a multi-adjoint lattice L, we consider an augmented language $\mathcal{L}_L^s \supseteq \mathcal{L}_L$ which may also include a number of symbolic values, symbolic adjoint pairs and symbolic connectives which do not belong to L. Symbolic objects are usually denoted as o^s with a superscript s and, in our online tool, their identifiers always start with #.

Definition 1 (sMALP *program*) Let L be a multi-adjoint lattice. An sMALP program over L is a set of symbolic rules, where each symbolic rule is a formula $(H \leftarrow_i \mathcal{B}$ *with* $v)$, where the following conditions hold:

- H is an atomic formula of $\mathcal{L}_L$ (the head of the rule);
- $\leftarrow_i$ is a (possibly symbolic) implication from either a symbolic adjoint pair $\langle \&^s, \leftarrow^s \rangle$ or from an adjoint pair of L;
- $\mathcal{B}$ (the body of the rule) is a symbolic goal, i.e., a well-formed formula of $\mathcal{L}_L^s$;
- v is either a truth degree (a value of L) or a symbolic value.

```
popularity(X) #<s1 facilities(X) #|s2 @aver(location(X),rates(X)) with 0.9.

facilities(sun) with #s3.     location(sun) with 0.4.    rates(sun) with 0.7.
facilities(sweet) with 0.5.   location(sweet) with 0.3.  rates(sweet).
facilities(lux) with 0.9.     location(lux) with 0.8.    rates(lux) with 0.2.
```

Fig. 3 Example of an sMALP program loaded into the FLOPER system

Example 1 Figure 3 displays an sMALP program. Here, we consider a travel agency
that offers booking services on three hotels, named *sun*, *sweet* and *lux*, where each one
of them is featured by three factors: the hotel facilities, the convenience of its location,
and the rates, denoted by predicates *facilities*, *location* and *rates*, respectively. Here,
we assume that all weights can be easily obtained except for the weight of the fact
facilities(*sun*), which is unknown, so we introduce a symbolic weight #s3. Also,
the programmer has some doubts on the connectives used in the first rule, so she
introduces two symbolic connectives, i.e., the implication and disjunction symbols
< s1 and # | s2.

3 Running and Unfolding Symbolic Programs

The procedural semantics of sMALP is defined in a stepwise manner as follows.
First, an *operational* stage –based on *admissible steps*, as described in Definition 2–
is introduced which proceeds similarly to SLD resolution in pure logic programming.
In contrast to standard logic programming, though, our operational stage returns an
expression still containing a number of (possibly symbolic) values and connectives.
Then, an *interpretive* stage –based on *interpretive steps* according to Definition 4–
evaluates these connectives and produces a final answer possibly containing symbolic values and connectives. The procedural semantics of both MALP and sMALP
programs is based on a similar scheme. The main difference is that, for MALP programs, the interpretive stage always returns a value, while for sMALP programs we
might get an expression containing symbolic values and connectives that should be
first instantiated in order to compute a final value.

In the following, $C[A]$ denotes a formula where A is a sub-expression which occurs
in the—possibly empty—context $C[]$. Moreover, $C[A/A']$ means the replacement of
A by A' in context $C[]$, whereas $Var(s)$ refers to the set of distinct variables occurring
in the syntactic object s, and $\theta[Var(s)]$ denotes the substitution obtained from θ by
restricting its domain to $Var(s)$. An sMALP *state* has the form $\langle Q; \sigma \rangle$ where Q is a
symbolic goal and σ is a substitution. We let $\mathcal{E}^s$ denote the set of all possible sMALP
states.

Definition 2 (*admissible step*) Let L be a multi-adjoint lattice and $\mathcal{P}$ an sMALP
program over L. An *admissible step* is formalized as a state transition system, whose

transition relation $\to_{AS} \subseteq (\mathcal{E}^s \times \mathcal{E}^s)$ is the smallest relation satisfying the following transition rules[1]:

1. $\langle \mathcal{Q}[A]; \sigma \rangle \to_{AS} \langle (\mathcal{Q}[A/v \&_i \mathcal{B}])\theta; \sigma\theta \rangle$,
   if $\theta = mgu(\{H = A\}) \neq fail$, $(H \leftarrow_i \mathcal{B} \; with \; v) \ll \mathcal{P}$ and $\mathcal{B}$ is not empty.[2]
2. $\langle \mathcal{Q}[A]; \sigma \rangle \to_{AS} \langle (\mathcal{Q}[A/\bot]); \sigma \rangle$,
   if there is no rule $(H \leftarrow_i \mathcal{B} \; with \; v) \ll \mathcal{P}$ such that $mgu(\{H = A\}) \neq fail$.

Here, $(H \leftarrow_i \mathcal{B} \; with \; v) \ll \mathcal{P}$ denotes that $(H \leftarrow_i \mathcal{B} \; with \; v)$ is a renamed apart variant of a rule in $\mathcal{P}$ (i.e., all its variables are fresh). Note that symbolic values and connectives are not renamed.

Observe that the second rule is needed to cope with expressions like $@_{aver}(p(a), 0.8)$, which can be evaluated successfully even when there is no rule matching $p(a)$ since $@_{aver}(0, 0.8) = 0.4$. We sometimes call *failure steps* to this kind of admissible steps because a pure logic language like **Prolog** would fail on such situations.

Given a relation $\to$ we denote by $\to^*$ its reflexive and transitive closure. Also, an *L^s-expression* is now a well-formed formula of $\mathcal{L}_L^s$ which is composed by values and connectives from L as well as by symbolic values and connectives.

Definition 3 (*admissible derivation*) Let L be a multi-adjoint lattice and $\mathcal{P}$ be an **sMALP** program over L. Given a goal $\mathcal{Q}$, an *admissible derivation* is a sequence $\langle \mathcal{Q}; id \rangle \to_{AS}^* \langle \mathcal{Q}'; \theta \rangle$. When $\mathcal{Q}'$ is an *L^s-expression*, the derivation is called *final* and the pair $\langle \mathcal{Q}'; \sigma \rangle$, where $\sigma = \theta[Var(\mathcal{Q})]$, is called a *symbolic admissible computed answer* (**saca**, for short) for goal $\mathcal{Q}$ in $\mathcal{P}$.

Example 2 Consider again the multi-adjoint lattice L and the **sMALP** program $\mathcal{P}$ of Example 1. Here, we have the following final admissible derivation for goal *popularity(X)* in $\mathcal{P}$ (the selected atom is underlined):

$\langle popularity(X); \; id \rangle$ $\to_{AS}$
$\langle \#\&s1(0.9, \#|s2(\underline{facilities(X)}, @aver(location(X), rates(X)))); \{X_1/X\} \rangle$ $\to_{AS}$
$\langle \#\&s1(0.9, \#|s2(\#s3, @aver(\underline{location(sun)}, rates(sun)))); \{X/sun, X_1/sun\} \rangle$ $\to_{AS}$
$\langle \#\&s1(0.9, \#|s2(\#s3, @aver(0.4, \underline{rates(sun)}))); \{X/sun, X_1/sun\} \rangle$ $\to_{AS}$
$\langle \#\&s1(0.9, \#|s2(\#s3, @aver(0.4, 0.7))); \{X/sun, X_1/sun\} \rangle$

Hence, the associated **saca** is $\langle \#\&s1(0.9, \#|s2(\#s3, @aver(0.4, 0.7))); \{X/sun\} \rangle$.

Given a goal $\mathcal{Q}$ and a final admissible derivation $\langle \mathcal{Q}; id \rangle \to_{AS}^* \langle \mathcal{Q}'; \sigma \rangle$, we have that $\mathcal{Q}'$ does not contain atomic formulas. Now, $\mathcal{Q}'$ can be *solved* by using the following interpretive stage:

---

[1]Here, we assume that A in $\mathcal{Q}[A]$ is the selected atom. Furthermore, as it is common practice, $mgu(E)$ denotes the *most general unifier* of the set of equations E [8].

[2]For simplicity, we consider that facts $(H \; with \; v)$ are seen as rules of the form $(H \leftarrow_i \top \; with \; v)$ for some implication $\leftarrow_i$. Furthermore, in this case, we directly derive the state $\langle (\mathcal{Q}[A/v])\theta; \sigma\theta \rangle$ since $v \&_i \top = v$ for all $\&_i$.

Definition 4 (*interpretive step*) Let L be a multi-adjoint lattice and $\mathcal{P}$ be an sMALP program over L. Given a saca $\langle Q; \sigma \rangle$, the *interpretive* stage is formalized by means of the following transition relation $\rightarrow_{IS} \subseteq (\mathcal{E}^s \times \mathcal{E}^s)$, which is defined as the least transition relation satisfying:

$$\langle Q[\zeta(r_1, \ldots, r_n)]; \sigma \rangle \rightarrow_{IS} \langle Q[\zeta(r_1, \ldots, r_n)/r_{n+1}]; \sigma \rangle$$

where ζ denotes a connective defined on L and $[\![\zeta]\!](r_1, \ldots, r_n) = r_{n+1}$.

An interpretive derivation of the form $\langle Q; \sigma \rangle \rightarrow_{IS}^* \langle Q'; \theta \rangle$ such that $\langle Q'; \theta \rangle$ cannot be further reduced, is called a *final* interpretive derivation. In this case, $\langle Q'; \theta \rangle$ is called a *symbolic fuzzy computed answer* (sfca, for short). Also, if Q' is a value of L, we say that $\langle Q'; \theta \rangle$ is a fuzzy computed answer (fca, for short).

Example 3 Given the saca of Example 2, we have the following final interpretive derivation (the connective reduced is underlined):

$$\langle \#\&s1(0.9, \#|s2(\#s3, @aver(0.4, 0.7))); \{X/sun\} \rangle \rightarrow_{IS}$$
$$\langle \#\&s1(0.9, \#|s2(\#s3, \underline{0.55})); \{X/sun\} \rangle$$

with $[\![@_{\text{aver}}]\!](0.4, 0.7) = 0.55$. Therefore, $\langle \#\&s1(0.9, \#|s2(\#s3, 0.55)); \{X/sun\} \rangle$ is a sfca of $popularity(X)$ in $\mathcal{P}$ since it cannot be further reduced.

On the other hand, *unfolding* is a well-known, widely used, semantics-preserving program transformation rule. The fold/unfold transformation approach was first introduced in [2] to optimize functional programs and then used for logic programs [18]. In essence, unfolding is usually based on the application of operational steps on the body of program rules [17]. The unfolding transformation is able to improve programs, generating more efficient code. Unfolding is the basis for developing sophisticated and powerful programming tools, such as fold/unfold transformation systems or partial evaluators, etc. Although in [4, 5] we successfully adapted such operation to MALP programs, there are not precedents coping with its symbolic extension, which motivates the present work.

Definition 5 (*Symbolic Unfolding*) Let $\mathcal{P}$ be an sMALP program and let $R : A \leftarrow \mathcal{B} \in \mathcal{P}$ be a program rule with no empty body. Then, the symbolic unfolding of rule R in program $\mathcal{P}$ is the new sMALP program $\mathcal{P}' = (\mathcal{P} - \{R\}) \cup \{A\sigma \leftarrow \mathcal{B}' \mid \langle \mathcal{B}; id \rangle \rightarrow \langle \mathcal{B}'; \sigma \rangle\}$.

Example 4 Considering again the sMALP program of Example 1, if we unfold its first rule (with selected atom *facilities*(X)) by applying a $\rightarrow_{AS}$ step with the three facts defining predicate *facilities*, we replace it with these three new rules:

popularity(*sun*)	$\# \leftarrow s1$ $\#|s2(\#s3, @_{\text{aver}}(location(sun), rates(sun)))$	*with* 0.9
popularity(*sweet*)	$\# \leftarrow s1$ $\#|s2(0.5, @_{\text{aver}}(location(sweet), rates(sweet)))$	*with* 0.9
popularity(*lux*)	$\# \leftarrow s1$ $\#|s2(0.9, @_{\text{aver}}(location(lux), rates(lux)))$	*with* 0.9

and it is easy to see that the new program produces the same set of s.f.c.a.'s for a given goal but reducing the length of derivations. We are nowadays identifying a set of sufficient conditions allowing us to prove the soundness and completeness properties of symbolic unfolding.

4 Conclusions and Future Work

In this work we have focused on a preliminary formulation of an unfolding transformation for optimizing sMALP programs. In contrast with other precedent fuzzy languages like MALP, the treatment of symbolic constants added to this extended framework introduces several risks for preserving the correctness of the transformation that we are proving nowadays. Moreover, since the sMALP language was initially conceived for tuning fuzzy logic programs, in the future we plan to explore the synergies between this technique and the unfolding transformation described in this paper. Finally, we also wish to manage similarity relations as recently done in [12] with the FASILL language (which represents a non-symbolic extension of MALP using unification by similarity [6, 7]).

References

1. Baldwin, J.F., Martin, T.P., Pilsworth, B.W.: Fril-Fuzzy and Evidential Reasoning in Artificial Intelligence. Wiley (1995)
2. Burstall, R.M., Darlington, J.: A transformation system for developing recursive programs. J. ACM **24**(1), 44–67 (1977)
3. Ishizuka, M., Kanai, N.: Prolog-ELF incorporating fuzzy logic. In: Aravind, K.J. (ed.) Proceedings of the 9th International Joint Conference on Artificial Intelligence, IJCAI'85, pp. 701–703. Morgan Kaufmann (1985)
4. Julián-Iranzo, P., Moreno, G., Penabad, J.: On fuzzy unfolding: a multi-adjoint approach. Fuzzy Sets Syst. **154**, 16–33 (2005)
5. Julián-Iranzo, P., Moreno, G., Penabad, J.: Operational/Interpretive unfolding of multi-adjoint logic programs. J. Univ. Comput. Sci. **12**(11), 1679–1699 (2006)
6. Julián-Iranzo, P., Moreno, G., Penabad, J.: Thresholded semantic framework for a fully integrated fuzzy logic language. J. Log. Algebra. Method Program. **93**, 42–67 (2017)
7. Julián-Iranzo, P., Moreno, G., Penabad, J., Vázquez, C.: A declarative semantics for a fuzzy logic language managing similarities and truth degrees. In: Proceedings of the 10th International Symposium on Rule Technologies Research, Tools, and Applications, RuleML 2016, Stony Brook, NY, USA, 6–9 July 2016. LNCS 9718, pp. 68–82. Springer (2016)
8. Lassez, J.L., Maher, M.J., Marriott, K.: Unification revisited. In: Minker, J. (ed.) Foundations of Deductive Databases and Logic Programming, pp. 587–625. Morgan Kaufmann, Los Altos, Ca. (1988)
9. Lee, R.C.T.: Fuzzy logic and the resolution principle. J. ACM **19**(1), 119–129 (1972)
10. Li, D., Liu, D.: A fuzzy Prolog Database System. Wiley (1990)
11. Medina, J., Ojeda-Aciego, M., Vojtáš, P.: Similarity-based Unification: a multi-adjoint approach. Fuzzy Sets Syst. **146**, 43–62 (2004)

12. Moreno, G., Penabad, J., Riaza, J.A.: On similarity-based unfolding. In: Proceedings of the 11th International Conference on Scalable Uncertainty Management, SUM'17, Granada, Spain, 4–6 Oct 2017. LNCS 10564, pp. 420–426. Springer (2017)
13. Moreno, G., Penabad, J., Riaza, J.A., Vidal, G.: Symbolic execution and thresholding for efficiently tuning fuzzy logic programs. In: Proceedings of the 26th International Symposium on Logic-Based Program Synthesis and Transformation, LOPSTR'16, Edinburgh, UK, 6–8 Sept 2016, Revised Selected Papers. LNCS 10184, pp. 131–147. Springer (2016)
14. Moreno, G., Penabad, J., Vidal, G.: Tuning fuzzy logic programs with symbolic execution (2016). CoRR, abs arXiv:1608.04688
15. Moreno, G., Riaza, J.A.: An online tool for tuning fuzzy logic programs. In: Proceedings of the International Joint Conference on Rules and Reasoning, RuleML+RR 2017, London, UK, 12–15 July 2017. LNCS 10364, pp. 184–198. Springer (2017)
16. Nguyen, H.T., Walker, E.A.: A First Course in Fuzzy Logic. Chapman & Hall, Boca Ratón, Florida (2006)
17. Pettorossi, A., Proietti, M.: Rules and strategies for transforming functional and logic programs. ACM Comput. Surv. 28(2), 360–414 (1996)
18. Tamaki, H., Sato, T.: Unfold/fold transformations of logic programs. In: Tärnlund, S. (ed.) Proceedings of Second Int'l Conference on Logic Programming, pp. 127–139 (1984)

Towards the Use of Hypergraphs in Multi-adjoint Logic Programming

Juan Carlos Díaz-Moreno, Jesús Medina and José R. Portillo

Abstract The representation of a logic program by a graph is a useful procedure in order to obtain interesting properties of the program and in the computation of the least model, when it exists. In this paper, we consider hypergraphs for representing multi-adjoint logic programs and, based on this representation, the hypotheses of an interesting termination result have been weakened.

1 Introduction

One of the most important problems in logic programming with non-decreasing operators is the computation of the least model of a given program. In order to obtain such a model the fix-point semantics is usually considered. This semantics is based on the iteration of the immediate consequence operator from the least interpretation. Since this iteration can be infinite, one important goal is to get termination properties of this iteration. In [3, 4], different termination theorems were introduced in the multi-adjoint logic programming. This logic programming framework was introduced in [10] as a generalization of different non-classical logic programming frameworks, such as the residuated logic programming [5] and the fuzzy logic programming framework presented in [11].

This paper considers directed hypergraphs [1, 8] in order to represent a multi-adjoint logic program and, based on this representation, introduce a termination result, which generalizes one of the most important termination theorems given in [4].

J. C. Díaz-Moreno · J. Medina
Department of Mathematics, University of Cádiz, Cádiz, Spain
e-mail: juancarlos.diaz@uca.es

J. Medina
e-mail: jesus.medina@uca.es

J. R. Portillo (✉)
Department of Applied Mathematics I, University of Sevilla, Sevilla, Spain
e-mail: josera@us.es

© Springer Nature Switzerland AG 2019
M. E. Cornejo et al. (eds.), *Trends in Mathematics and Computational Intelligence*, Studies in Computational Intelligence 796,
https://doi.org/10.1007/978-3-030-00485-9_6

2 Basic Definitions on Hypergraphs

This section recalls the notions we will need throughout the paper related to hypergraphs. For basic notions of graph theory see [2].

A *graph* is a pair of sets (V, E). V is the set of *vertices* or *nodes*. E is a set of 2-element subsets of V, named *edges*. The edges may be directed or undirected (the pairs are ordered or not). Directed edges are called *arcs*. A *cycle* in a graph is a path of edges and vertices wherein a vertex is reachable from itself. I.e., a ordered set of vertices $\{u_1, \ldots, u_i, \ldots u_p\}$ such that $u_i u_{i+1}$ is an edge of the graph and $u_1 = u_p$.

The first notion for *hypergraph*s is the definition, which is a generalization of a graph in which an edge is a non-empty subset of vertices. Specifically, a hypergraph $\mathcal{H}$ is a pair $\mathcal{H} = (V, E)$ where V is a set of elements called *nodes* or *vertices*, and E is a set of non-empty subsets of V called *hyperedges* or edges, see [2] for more details. Therefore, E is a subset of $\mathcal{P}(V) \setminus \{\emptyset\}$, where $\mathcal{P}(V)$ is the power set of V. Note that, when the cardinal of all hyperedges is 2, the hypergraph is a standard graph.

The generalization of a directed graph is called directed hypergraph and contains directed hyperedges. A *directed hyperedge* or *hyperarc* is an ordered pair, $e = (X, Y)$, of (possibly empty) disjoint subsets of vertices; X is called the *tail* of e and Y is its *head*. From now on, the tail and the head of an hyperarc e will be denoted by $T(e)$ and $H(e)$, respectively.

Hence, a directed hypergraph is a hypergraph with directed hyperedges [1, 8]. A *backward hyperarc*, or simply *B-arc*, is a hyperarc $e = (T(e), H(e))$, where the head exactly has one vertex. When all the hyperarcs of a hypergraph are B-arcs, then the hypergraph is called *B-graph* (or *B-hypergraph*) [8]. For example, the hypergraph $\mathcal{H} = (V, E)$ introduced on the left of Fig. 1, where $V = \{a, b, c, d\}$ and $E = \{(\{a\}, \{b\}), (\{b, c\}, \{a\}), (\{c, d\}, \{a\})\}$ is a B-graph. This paper will only consider this kind of hypergraphs.

B-graphs (and the dually defined F-graphs) are a useful tool in different applications [1, 8, 9]. As a consequence, they have been introduced many times in the literature with various names. For example, the labelled graphs, used in [6, 7] to represent Horn formulae, are B-graphs.

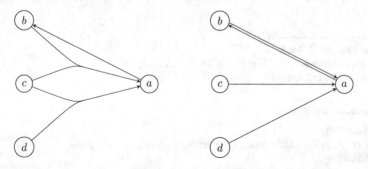

Fig. 1 Left: Example of B-graph $\mathcal{H} = (V, E)$. Right: Directed graph subjacent to $\mathcal{H}$

In contrast with ordinary graphs for which there is a single natural notion of cycles and acyclic graphs, there are multiple natural non-equivalent definitions of acyclicity for hypergraphs which collapse to ordinary graph acyclicity for the special case of ordinary graphs.

In this paper, we only need to consider the cycles on the subjacent directed graph of a B-graph. Given any directed hypergraph $\mathcal{H} = (V, E)$, the subjacent directed graph $G(\mathcal{H}) = (V(G), E(G))$ has the same nodes that $\mathcal{H}$, i.e. $V(G) = V$ and an arc exists in $E(G)$ from the node u to the node v if and only if it exists a hyperedge $e \in E$ such that $u \in T(e)$ and $v \in H(e)$. For example, Fig. 1 shows on the right the subjacent graph to the hypergraph on the left. The vertices a and b form a cycle in that graph, but the hypergraph has not hypercycles under the common definitions [2].

3 Multi-adjoint Logic Programming

This section recalls the algebraic structure considered in this framework, the notion of multi-adjoint logic program, and one of the most interesting termination theorems introduced in [4]. The basic operators considered in this framework are adjoint pairs.

Definition 1 Given a partially ordered set $(P, \leq)$, the pair $(\&, \leftarrow)$ is an *adjoint pair* with respect to $(P, \leq)$ if the mappings $\&, \leftarrow: P \times P \rightarrow P$ satisfy that:

1. $\&$ is order-preserving in both arguments.
2. $\leftarrow$ is order-preserving in the first argument (the consequent) and order-reversing in the second argument (the antecedent).
3. The equivalence $x \leq y \leftarrow z$ if and only if $x \& z \leq y$ holds, for all $x, y, z \in P$.

The algebraic structure considered in this logic programming framework is called multi-adjoint lattice.

Definition 2 A *multi-adjoint lattice* is a tuple $(L, \preceq, \leftarrow_1, \&_1, \ldots, \leftarrow_n, \&_n)$ verifying the following properties:

1. $(L, \preceq)$ is bounded lattice, i.e. it has bottom ($\perp$) and top ($\top$) elements;
2. $(\&_i, \leftarrow_i)$ is an adjoint pair in $(L, \preceq)$, for all $i \in \{1, \ldots, n\}$;
3. $\top \&_i \vartheta = \vartheta \&_i \top = \vartheta$, for all $\vartheta \in L$ and for all $i \in \{1, \ldots, n\}$.

Given a multi-adjoint lattice, a set of propositional symbols Π, a given language denoted as $\mathfrak{F}$ and different monotonic operators defined on L, the notion of program (set of rules) is introduced in this framework.

Definition 3 Given a multi-adjoint lattice $(L, \preceq, \leftarrow_1, \&_1, \ldots, \leftarrow_n, \&_n)$. A *multi-adjoint logic program* $\mathbb{P}$ is a set of rules of the form $\langle (A \leftarrow_i B), \vartheta \rangle$ such that:

1. The *rule* $(A \leftarrow_i \mathcal{B})$ is a formula of $\mathfrak{F}$;
2. The *confidence factor* ϑ is an element (a truth-value) of L;
3. The *head* of the rule A is a propositional symbol of Π.
4. The *body* formula $\mathcal{B}$ is a formula of $\mathfrak{F}$ built from propositional symbols $B_1, \ldots, B_n$ $(n \geq 0)$ by the use of conjunctors $\&_1, \ldots, \&_n$ and $\wedge_1, \ldots, \wedge_k$, disjunctors $\vee_1, \ldots, \vee_l$, aggregators $@_1, \ldots, @_m$ and elements of L.
5. *Facts* are rules with body $\top$.

This paper will be focused on one of the most important theorems introduced in [4]. Before recalling this result we need different definitions.

Definition 4 Let $\mathbb{P}$ be a multi-adjoint program, and $A \in \Pi$. The set $R_{\mathbb{P}}^I(A)$ of *relevant values* for A with respect to an interpretation I is the set of maximal values of the set $\{\vartheta \&_i \hat{I}(\mathcal{B}) \mid \langle A \leftarrow_i \mathcal{B}, \vartheta \rangle \in \mathbb{P}\}$.

The immediate consequences operator, given by van Emden and Kowalski, is defined in this framework as follows.

Definition 5 Given a multi-adjoint logic program $\mathbb{P}$. The *immediate consequences operator* $T_{\mathbb{P}}$ maps interpretations to interpretations, and for an interpretation I and an arbitrary propositional symbol A is defined by

$$T_{\mathbb{P}}(I)(A) = \sup\{\vartheta \&_i \hat{I}(\mathcal{B}) \mid \langle A \leftarrow_i \mathcal{B}, \vartheta \rangle \in \mathbb{P}\}$$

The main feature of $T_{\mathbb{P}}$ is that its least fix-point coincides with the least model of the program $\mathbb{P}$ [10]. Since the least fix-point is computed iterating the $T_{\mathbb{P}}$ operator from the least interpretation, Δ, it is important to know when this iteration finishes in a finite number of steps.

The termination theorem in [4] was introduced for sorted and local multi-adjoint logic programs. In order to simplify the notation, we have adapted it for (uni-sorted) multi-adjoint logic programs.

Theorem 1 *Given a multi-adjoint logic program $\mathbb{P}$ with finite dependences and where the operators $@ : L^m \to L$ in the body of the rules satisfy the boundary condition with the $\top$ element, that is,*

$$@(\underbrace{\top, \ldots, \top}_{k}, x, \underbrace{\top, \ldots, \top}_{m-k-1}) \preceq x$$

for all $x \in L$. If for every iteration n and propositional symbol A the set of relevant values for A with respect to $T_{\mathbb{P}}^n(\Delta)$ is a singleton, then $T_{\mathbb{P}}$ terminates for every query.

This result will be weakened in the following section.

4 Representing Programs by Hypergraphs

This section presents a simple example which does not satisfy the hypotheses of Theorem 1, but the least model of the program is obtained after finitely many iterations. Then, in order to extend this result to a bigger number of programs, a straightforward mechanism for representing a logic program by a B-graph is introduced. Finally, based on this representation, the hypotheses of Theorem 1 will be weakened.

Example 1 Consider the following program $\mathbb{P}$:

$$\langle a \leftarrow_P b \,\&_G\, c, 0.8 \rangle \qquad\qquad \langle b \leftarrow_P a, 0.7 \rangle$$
$$\langle a \leftarrow_P @(d, c), 1.0 \rangle \qquad\qquad \langle c \leftarrow_P 1.0, 1.0 \rangle$$
$$\langle a \leftarrow_P 1.0, 0.5 \rangle$$

where the aggregator $@: [0, 1] \times [0, 1] \to [0, 1]$ is the weighted sum defined as $@(x, y) = (x + 3y)/4$, for all $x, y \in [0, 1]$.

Although the minimum operator $\&_G$ satisfies the boundary condition with the 1 element (hypothesis in Theorem 1), the aggregator @ does not verify it and so, we cannot apply this theorem in order to know whether the computation of the least model terminates in a finite number of iterations. However, in this case, only 3 iterations are needed, as we show below:

	a	b	c	d
$T_{\mathbb{P}}^0 =$	0.0	0.0	0.0	0.0
$T_{\mathbb{P}}^1 =$	0.75	0.0	1.0	0.0
$T_{\mathbb{P}}^2 =$	0.75	0.525	1.0	0.0
$T_{\mathbb{P}}^3 =$	0.75	0.525	1.0	0.0

This example shows that the hypotheses in Theorem 1 should be weakened. For that, we will represent a program by a B-graph and we will relate the termination of the iterations to the existence of cycles in the subjacent directed graph and whether aggregator operators, which do not satisfy the hypotheses of the theorem, are involved in these cycles.

Note that, it has been possible to compute the least model in a finite number of iterations because the aggregator operator @ is not in a cycle of the subjacent digraph of the associated B-graph.

The associated B-graph $\mathcal{H}_{\mathbb{P}}$ associated with a program $\mathbb{P}$ is constructed as follows: the vertex set of the hypergraph is the propositional symbol set Π of the program. Hence, for the program $\mathbb{P}$ in Example 1, we have $V(\mathcal{H}_{\mathbb{P}}) = \{a, b, c, d\}$. One hyperarc will be obtained from each rule as follows: Given a rule, the propositional symbols of the antecedent of the rule will be the tail $T(e)$ of the associated hyperarc and the propositional symbol of the head of the rule will be the only element of the head $H(e)$ of the associated hyperarc. This hyperarc is labeled with the aggregator in the body of the rule. When no aggregator operator appears in the body of the rule, we will

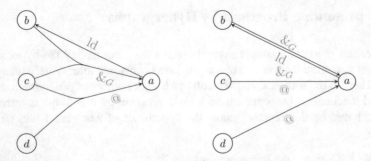

Fig. 2 Left: (Labelled) B-graph associated with the program given in Example 1. Right: (Labelled) Subjacent directed graph from the B-graph on the left

consider the identity mapping. For example, from the rule $\langle a \leftarrow_P b \&_G c, 0.8 \rangle$ for the program $\mathbb{P}$ in Example 1, we obtain the hyperarc $(\{b, c\}, \{a\})$ with the label $\&_G$. Due to the considered mechanism, the hypergraph resultant is always a (labelled) B-graph. Figure 2 (left) shows the associated B-graph of the program $\mathbb{P}$ given in Example 1.

Finally, a weak version of Theorem 1 is introduced:

Theorem 2 *Given a multi-adjoint logic program $\mathbb{P}$ with finite dependences and the B-graph $\mathcal{H}_\mathbb{P}$ associated with $\mathbb{P}$. If the operators involved in the cycles of the subjacent directed graph of $\mathcal{H}_\mathbb{P}$ satisfy the boundary condition with the $\top$ element and, for every iteration n and propositional symbol A, the set of relevant values for A with respect to $T_\mathbb{P}^n(\triangle)$ is a singleton, then $T_\mathbb{P}$ terminates for every query.*

Note that this result only needs that the aggregators operators involved in the cycles of the subjacent directed graph of the B-graph associated with the program satisfy the boundary condition. Therefore, this result is notably more general than Theorem 1 and can be applied, for example, to the program given in Example 1.

5 Conclusions and Future Work

This paper has presented a procedure in order to represent a multi-adjoint logic program as a hypergraph. As a first consequence, we have generalized one of the most important termination results introduced in [4]. This representation will provide more interesting properties in the future. We will study other efficient termination results and we will analyze analogies between different notions in logic programming and in graph theory in order to create synergies between both theories. The obtained results will also be compared with the existent ones in the literature.

References

1. Ausiello, G., Laura, L.: Directed hypergraphs: introduction and fundamental algorithms—a survey. Theor. Comput. Sci. **658**, 293–306 (2017)
2. Berge, C.: Graphs and Hypergraphs. Elsevier Science Ltd. (1985)
3. Damásio, C., Medina, J., Ojeda-Aciego, M.: Sorted multi-adjoint logic programs: termination results and applications. In: Lecture Notes in Artificial Intelligence, vol. 3229, pp. 252–265 (2004)
4. Damásio, C., Medina, J., Ojeda-Aciego, M.: Termination of logic programs with imperfect information: applications and query procedure. J. Appl. Log. **5**, 435–458 (2007)
5. Damásio, C.V., Pereira, L.M.: Monotonic and residuated logic programs. In: Symbolic and Quantitative Approaches to Reasoning with Uncertainty, ECSQARU 2001. Lecture Notes in Artificial Intelligence, vol. 2143, pp. 748–759 (2001)
6. Dowling, W.F., Gallier, J.H.: Linear-time algorithms for testing the satisfiability of propositional Horn formulae. J. Log. Program. **1**(3), 267–284 (1984)
7. Gallo, G., Urbani, G.: Algorithms for testing the satisfiability of propositional formulae. J. Log. Program. **7**(1), 45–61 (1989)
8. Gallo, G., Longo, G., Pallottino, S., Nguyen, S.: Directed hypergraphs and applications. Discrete Appl. Math. **42**(2–3), 177–201 (1993)
9. Jeroslow, R.G., Martin, R.K., Rardin, R.L., Wang, J.: Gainfree Leontief substitution flow problems. Math. Program. **57**, 375–414 (1992)
10. Medina, J., Ojeda-Aciego, M., Vojtáš, P.: Multi-adjoint logic programming with continuous semantics. In: Lecture Notes in Artificial Intelligence, vol. 2173, pp. 351–364 (2001)
11. Vojtáš, P.: Fuzzy logic programming. Fuzzy Sets Syst. **124**(3), 361–370 (2001)

Banking Applications of FCM Models

Miklós F. Hatwágner⑩, Gyula Vastag, Vesa A. Niskanen
and László T. Kóczy

Abstract Fuzzy Cognitive Map (FCMs) is an appropriate tool to describe, qualitatively analyze or simulate the behavior of complex systems. FCMs are bipolar fuzzy graphs: their building blocks are the concepts and the arcs. Concepts represent the most important components of the system, the weighted arcs define the strength and direction of cause-effect relationships among them. FCMs are created by experts in several cases. Despite the best intention the models may contain subjective information even if it was created by multiple experts. An inaccurate model may lead to misleading results, therefore it should be further analyzed before usage. Our method is able to automatically modify the connection weights and to test the effect of these changes. This way the hidden behavior of the model and the most influencing concepts can be mapped. Using the results the experts may modify the original model in order to achieve their goal. In this paper the internal operation of a department of a bank is modeled by FCM. The authors show how the modification of the connection weights affect the operation of the institute. This way it is easier to understand the working of the bank, and the most threatening dangers of the system getting into an unstable (chaotic or cyclic state) can be identified and timely preparations become possible.

M. F. Hatwágner (✉) · L. T. Kóczy
Department of Information Technology, Széchenyi István University, Győr, Hungary
e-mail: miklos.hatwagner@sze.hu

L. T. Kóczy
e-mail: koczy@sze.hu; koczy@tmit.bme.hu

G. Vastag
Department of Leadership and Organizational Communication, Széchenyi István University,
Győr, Hungary
e-mail: vastag.gyula@sze.hu

V. A. Niskanen
Department of Economics and Management, University of Helsinki, Helsinki, Finland
e-mail: vesa.a.niskanen@helsinki.fi

L. T. Kóczy
Department of Telecommunications and Media Informatics, Budapest University of Technology
and Economics, Budapest, Hungary

© Springer Nature Switzerland AG 2019
M. E. Cornejo et al. (eds.), *Trends in Mathematics and Computational
Intelligence*, Studies in Computational Intelligence 796,
https://doi.org/10.1007/978-3-030-00485-9_7

61

Keywords Banking · Fuzzy cognitive maps · Model uncertainty
Multiobjective optimization · Bacterial evolutionary algorithm

1 Introduction

Fuzzy Cognitive Maps (FCM) are suitable to describe complex systems for decision makers. The models include the most important system components and the direction and strength of relationships among them. There are numerous papers in the literature dealing with how to establish models and how to perform simulations with them to support decision making tasks (e.g. [1]). The internal operation of a bank is described and analyzed in this paper with FCM. The behavioral uncertainty and stability of the model were also investigated. The applied method [2] examines the effect of small changes of weight on model behavior. At least two good reasons exists why such an analysis is worth performing.

The original model was provided by experts, and according to our experience the connection matrix provided by humans is sometimes not perfect. It is not surprising that it is sometimes not easy to define the weight of a connection between two components. Even if they do their best, the number of connections is commensurate with the square of the number of components. In our case the investigated system had 13 components, thus the number of connections can theoretically be up to 156. In general, it is very hard to see the investigated system as a whole with all its details and to choose appropriate weights to represent the real relationships well. If these weights are not properly estimated, the simulation of the system will lead to outcomes that may never occur under real circumstances.

Even if the weights are defined properly the results of this paper may be interesting, because this way a more complete understanding of the model behavior can be obtained. This knowledge may lead to a modified model that eventuates better operation, helps exploring the effects that may jeopardize the operation of the system, etc.

The next section describes briefly the theoretical background of the applied methods, including FCMs in general and the method of uncertainty analysis which is used to find the most interesting, slightly modified model versions. It is followed by the analysis and modeling results of the banking system. Finally, the directions of further improvements and conclusions are summarized.

2 A Short Overview of Fuzzy Cognitive Maps

Axelrod [3] suggested first the use of cognitive maps to support decision making in politics. This idea was later extended to Fuzzy Cognitive Maps (FCMs) by Kosko [4, 5]. FCMs are directed, bipolar fuzzy graphs [6]. The nodes of this structure represent the main components of a system, and are usually called 'concepts'. Their values,

which fall into the unit interval [0, 1] [7], express the current state of a component
(e.g. a partially opened tap) [8]. The arcs among concepts represent the relationships
in the system. The weights assigned to the arcs falls in the [− 1, 1] interval, where
the sign of weights define the direction (amplifying, suppressing), and the absolute
value of it the strength of the connection.

FCMs are often visualized by a graph or described by the connection matrix which
contains the weights of arcs. According to Kosko's original idea, self-loops are not
allowed, therefore the main diagonal of the matrix contains only zeros.

The most important capability of FCMs is that simulations can be performed with
them in order to predict the future states of the system. If the initial values of concepts
are known and also the connection weights are given, the next state of the model can
be calculated by Eq. (1).

$$A_i^{(t+1)} = f\left(\sum_{j=1}^{M} w_{ji} A_j^{(t)} + A_i^{(t)} \right), i \neq j \tag{1}$$

In the equation $A_i^{(t)}$ represents the value of concept i at time t (also called 'activation' value), w_{ji} is the weight of the directed arc between concepts j and i, M is
the number of concepts and f is the threshold function.

We note here that several alternative equations are used besides (1), which alter-
natives were first proposed in [9]. This version was chosen in this paper because it
also uses the current value of the concept when calculating the next value. It means
that concepts have a 'memory', and it affects their future states. This behavior is very
common in real systems.

The role of the threshold function is to keep the activation values in the allowed
range. Several versions of this function are described in the literature [7], but only
one of them, the most popular one, namely the sigmoidal function (2) was applied.

$$f = \frac{1}{1 + e^{-\lambda x}} \tag{2}$$

The function has a λ parameter that defines the steepness. The value of $\lambda = 5$ is
often used in literature, and thus it was chosen by the authors as well. It must be
noted however that this parameter value may change the results considerably, and
the effect of different values should be further analyzed in the future. A simulation
may end up in three different ways [7]:

1. Generally the values of concepts converge to a final, stable value in a dozen
   discrete time steps. The final vectors of concept values are called the 'fixed point
   attractors'.

2. Sometimes a series of n state vectors appears repeatedly after a specific time step
    of the simulation, which is called a 'limit cycle'.
3. The last possible outcome is, when the values of concepts never stabilize, and
    the model behaves chaotically.

Generally, limit cycles and most of all chaotic behavior should be avoided, because
in these cases the future states of the system cannot be predicted. In some specific
applications however, e.g. if the goal is predict time series data [10], this behavior
can be useful.

3 Description of the Method Applied to Analyze the Uncertainty of Connection Weights

The main idea of uncertainty investigation is to modify the connection weights and
then to analyze the effect of modifications by simulations. The modifications are
directed by the Bacterial Evolutionary Algorithm (BEA) [11–13], in order to find
the most 'interesting' model variants: models with more fixed point attractors and/or
chaotic behavior. These outcomes are found here by starting simulations with the
same set of 1000 different, randomly generated initial state vectors, the so-called
scenarios.

BEA is an evolutionary algorithm which is able to find the quasi-optimum of
even a non-continuous, non-linear, multimodal function. It starts with a population
of possible solutions and improves these solution candidates (also called 'bacteria')
in every consecutive generation. The two main operators, bacterial mutation and
gene transfer help to achieve this goal. The first one explores the search space by the
random modifications of genetic data, the second one combines the already existing
genetic information of the population.

In this specific case, the bacteria of the BEA represent modified connection matrices. In our experiment the population consisted of 50 bacteria, and 5 consecutive generations were created. The weights in an FCM model are represented by real numbers,
thus their number is theoretically infinite. Obviously, the number of weights had to be
limited to a certain number, in our case 9 (-1, -0.75, -0.5, ..., 1). The investigated
model used only five discrete levels according to the linguistic variables chosen by
experts. The 9 levels made possible smoother changes in connection weights, and
according to our experience, the use of more levels does not provide significant
advantages. The concept values in scenarios were also limited to five discrete levels.

In order to limit the computational demand of the algorithm, and because human
experts can identify concepts without any connections with high confidence, the
elements of the connection matrix containing zeroes were left untouched.

Table 1 Concept IDs, names and categories of the investigated model

Concept ID	Concept name	Category
C1	Clients	Assets
C2	Rules and regulations	
C3	New IT solutions	
C4	Funding	Money
C5	Cost reduction	
C6	Profit/loss	Financials
C7	Investments	
C8	Staff	Human resources
C9	New services	Product and process development
C10	Quality	
C11	Client development	Output measures
C12	Service development	
C13	Productivity	

Despite of these restrictions it is easy to see that an exhaustive search would have been impossible in practice: the model under investigation contains 13 concepts, all concepts can have one of the possible 9 levels, thus the number of possible connection matrices can be up to 7.275e+148 (depending on the number of zero weight connections), the number of scenarios with five discrete levels is 1.22e+9. That is why BEA was applied to find the interesting modified models. The λ parameter of the threshold function was set to five in all simulations, because it would have further increased the execution time of the program, and simulations themselves can be time consuming tasks. Limit cycles and chaotic behavior cannot be distinguished by the program yet, but the fixed point attractors were recognized automatically.

4 Results

The model describes the components (concepts) of a bank playing the key roles in this research and their relations including their strength and direction. The concept id's, their corresponding names are collected in Table 1. The concepts can be categorized into six different groups. Table 2 contains the connection matrix of the model.

First, the original model was investigated by simulations. The sigmoid type threshold function was applied with $\lambda = 5$ steepness parameter. Using a thousand element, random-generated set of initial state vectors (scenarios), two possible outcomes were detected by the K-Means clustering method. Both of them were fixed-point attractors (Fps), and most of the concepts had the same final values (1.0), except C6 (Profit/loss) and C8 (Staff). We remark here, that C4 was an input concept and as such it did not change its value during simulations, but the specific value itself depended

Table 2 Connection matrix of the FCM model

	C1	C2	C3	C4	C5	C6	C7	C8	C9	C10	C11	C12	C13
C1	0	0	0.5	0	0	0.5	1	0.5	0	0.5	1	0.5	0
C2	1	0	0.5	1	0	0	1	1	0.5	0	1	1	0
C3	1	0.5	0	0	0	-1	0	-1	1	0	1	1	1
C4	0	0	0	0	0	0	0	0	0	0	0	0	0
C5	0	0	1	-0.5	0	0	0	-1	0	0	0	1	0
C6	0	0	0	0	-0.5	0	0	0	0	0	0	0	0
C7	0.5	0	0.5	1	0	0.5	0	0	0	-0.5	0	0	0
C8	0	0	0	0	0	-0.5	0	0	0	0.5	0	0	-0.5
C9	0	0	0	1	0	0.5	0.5	0.5	0	-0.5	1	0.5	0
C10	0.5	0	0	0	0	0	0.5	0.5	0.5	0	0	0	0
C11	0	0	0.5	0.5	0	0	0	0	0.5	0.5	0.5	0	1
C12	0	0	0.5	0.5	0	0	1	0	0.5	0	0	0	-0.5
C13	0	0	1	0	0	0.5	0	0	0	0	1	0	0

Table 3 Fixed-point attractors of the model

Concepts	C1–C3, C5, C7, C9–C13	C6	C8
FP#1	1.000	0.150	0.990
FP#2	1.000	0.855	0.922

on the content of the initial state vector only and was consequently left out from clustering. The final values of concepts are collected in Table 3. The first FP appeared in 23.1% of all investigated cases, and the second in the remaining 76.9%.

Next, the model was further analyzed to reveal the effect of modified connection weights on its behavior. The search directed by BEA found 50 interesting model variants, but considering the size limitations of the paper, only two of them are presented here. The connection matrix of the first variant is shown in Table 4. The values in parenthesis show the original connection weights to make comparisons easier. This modified model resulted in 12 different fixed-point attractors, but never behaved chaotically or produced limit cycles. The final state vectors are collected in Table 5.

Some interesting phenomena can be observed in Table 5. The value of C8 was very high in the original model (≈ 0.9), but it can be close to zero in the modified model. The values of C3, C11 and C13 were one, but in the modified model various values can be observed. The FP values of C2, C5, C7, C9 and C12 were 1 in the original model, it practically did not changed despite the modifications. C6 had two different values in the original model, but only a single one after the modifications. C10 changed its value from 1 to 0.

The second selected model variant example behaved in a different way: it had only 9 FPs, but 882 simulations out of 1000 did not result in stable state (chaotic behavior or limit cycles) (Tables 6 and 7).

The FP values of C1, C2, C5, C7, C9, C11 and C12 were exclusively 1, but in the modified model their values could be significantly different. The value of C3 was always 1 in the original model, and it practically did not change after the model modifications. C6 had two different values (a low and a high one) in case of both model versions, but these pairs of values are not the same. C8 had two high FP values in the original model, but hold two smaller values in the modified model. The FP values of C10 and C13 are still 1.

5 Conclusions and Future Research

The applied method generated small modifications on FCM models that led to very different model behaviors. It proved to be very useful to find relationships that are sensitive to changes and may cause unexpected simulation results. These connection weights need further investigations by experts of the specific field.

Table 4 Connection matrix of the first model variant

	C1	C2	C3	C4	C5	C6	C7	C8	C9	C10	C11	C12	C13
C1	0	0	−0.75 (0.5)	0	0	1 (0.5)	−1 (1)	−0.25 (0.5)	0	−0.5 (0.5)	0.5 (1)	0.75 (0.5)	0
C2	1	0	0.75 (0.5)	−0.5 (1)	0	0	0.5 (1)	−0.25 (1)	0 (0.5)	0	0.25 (1)	1	0
C3	0 (1)	0.75 (0.5)		0	0	−0.25 (−1)			−0.75 (1)	0	0.75 (1)	−0.75 (1)	1
C4	0	0	0	0	0	0	0	0	0	0	0	0	0
C5	0	0	−0.5 (1)	0 (−0.5)	0	0	0	1 (−1)	0	0	0	1	0
C6	0	0	0	0	−0.75 (−.5)	0	0	0	0	0	0	0	0
C7	1 (0.5)	0	1 (0.5)	−0.25 (1)	0	−1 (0.5)		0	0	−0.75 (−0.5)	0	0	0
C8	0	0	0	0	0	−1 (−0.5)	0	0	0	−0.25 (0.5)	0	0	−0.5
C9	0	0	0	1	0	−0.25 (0.5)	1 (0.5)	0.75 (0.5)	0	0.75 (−0.5)	0	−0.5 (0.5)	0
C10	−0.25 (0.5)	0	0 (0.5)	−0.75 (0.5)	0	0	−1 (0.5)	−0.25 (0.5)	−1 (0.5)	0	−0.75 (1)	0	0
C11	0	0	0	0	0	0	0	0	1 (0.5)	−1 (0.5)	0	0	−1 (1)
C12	0	0	−0.75 (0.5)	0.5	0	0	−0.5 (1)	0	0.25 (0.5)	0	0.25 (0.5)	0	0.75 (−.5)
C13	0	0	1	0	0	−0.75 (0.5)	0	0		0	−0.5 (1)	0	0

Table 5 Fixed-point attractors of the first model variant

FP ID	C1	C2	C3	C5	C6	C7	C8	C9	C10	C11	C12	C13
FP#1	0.826	1.000	1.000	1.000	0.026	1.000	0.112	1.000	0.000	0.982	0.995	0.999
FP#2	0.037	1.000	0.997	1.000	0.026	1.000	0.787	1.000	0.000	0.960	0.996	0.999
FP#3	0.108	1.000	1.000	1.000	0.026	1.000	0.112	1.000	0.000	0.943	0.997	0.999
FP#4	0.006	1.000	0.848	1.000	0.026	1.000	0.785	1.000	0.000	0.053	0.998	1.000
FP#5	0.008	1.000	0.997	1.000	0.026	1.000	0.112	1.000	0.000	0.046	0.998	1.000
FP#6	0.994	1.000	0.015	1.000	0.026	1.000	0.989	1.000	0.000	0.999	0.998	0.138
FP#7	0.188	1.000	0.109	1.000	0.026	0.994	0.796	1.000	0.000	0.036	1.000	0.995
FP#8	0.993	1.000	0.046	1.000	0.026	1.000	0.981	1.000	0.000	0.998	0.999	0.338
FP#9	0.026	1.000	0.488	1.000	0.026	0.998	0.788	1.000	0.000	0.065	1.000	0.999
FP#10	0.621	1.000	1.000	1.000	0.026	1.000	0.112	1.000	0.000	0.965	0.997	0.999
FP#11	0.392	1.000	0.098	1.000	0.026	0.998	0.799	1.000	0.000	0.027	1.000	0.995
FP#12	0.870	1.000	0.120	1.000	0.026	1.000	0.795	1.000	0.000	0.047	1.000	0.995

Table 6 Connection matrix of the second model variant

	C1	C2	C3	C4	C5	C6	C7	C8	C9	C10	C11	C12	C13
C1	0	0	0.75 (0.5)	0	0	-0.75 (0.5)	-0.25 (1)	0.25 (0.5)	0	-1 (0.5)	0.25 (1)	1 (0.5)	0
C2	-1 (1)	0	-0.75 (0.5)	-1 (1)	0	0	0.25 (1)	-1 (1)	0.75 (0.5)	0	0.25 (1)	-1 (1)	0
C3	-1 (1)	0.5	0	0	0	-0.25 (-1)	0	-1	-0.5 (1)	0	-0.75 (1)	-1 (1)	0.5 (1)
C4	0	0	0	0	0	0	0	0	0	0	0	0	0
C5	0	0	-0.25 (1)	0.5 (-0.5)	0	0	0	-0.75 (-1)	0	0	0	-0.5 (1)	0
C6	0	0	0		-0.75 (-.5)	0	0	0	0	0	0	0	0
C7	-0.25 (0.5)	0	-0.5 (0.5)	0.75 (1)	0	0.5	0	0	0	-1 (-0.5)	0	0	-1 (-0.5)
C8	0	0	0	0	0	0.75 (-0.5)	0	0	0	-0.25 (0.5)	0	0	0
C9	0	0	0	-0.75 (1)	0	-0.5 (0.5)	0.25 (0.5)	0 (0.5)	0	0 (-0.5)	0	-0.5 (0.5)	0
C10	0.5	0	0	0	0	0	1 (0.5)	0.75 (0.5)	0.5	0	-0.25 (1)	0	0
C11	0	0	0.25 (0.5)	-0.25 (0.5)	0	0	0	0	1 (0.5)	-0.25 (0.5)	0	0	0 (1)
C12	0	0	1 (0.5)	-0.75 (0.5)	0	0	0.5 (1)	0	0 (0.5)	0	-0.75 (0.5)	0	0 (-0.5)
C13	0	0	0.5 (1)	0	0	0.75 (0.5)	0	0	0	0	0.75 (1)	0	0

Table 7 Fixed-point attractors of the second model variant

FP ID	C1	C2	C3	C5	C6	C7	C8	C9	C10	C11	C12	C13
FP#1	0.053	0.000	0.000	0.010	0.993	0.957	0.846	0.012	1.000	0.964	0.883	1.000
FP#2	0.001	0.020	0.001	0.201	0.985	0.979	0.834	0.065	1.000	0.968	0.103	1.000
FP#3	0.970	0.000	0.000	0.994	0.027	0.883	0.002	0.007	1.000	0.158	0.923	0.997
FP#4	0.002	0.280	0.001	0.046	0.992	0.137	0.844	0.182	1.000	0.990	0.041	1.000
FP#5	0.002	0.387	0.006	0.045	0.992	0.134	0.844	0.180	1.000	0.990	0.041	1.000
FP#6	0.002	0.995	0.001	0.046	0.992	0.137	0.844	0.880	1.000	1.000	0.039	1.000
FP#7	0.013	0.001	0.000	0.024	0.992	0.979	0.845	0.013	1.000	0.957	0.656	1.000
FP#8	0.033	0.044	0.000	0.073	0.991	0.829	0.135	0.023	1.000	0.977	0.961	1.000
FP#9	0.020	0.000	0.000	0.061	0.991	0.998	0.843	0.001	1.000	0.146	0.950	1.000

The method should be further improved, however. The extent of weight modifications should be limited to a certain degree, depending on the application area. The effect of modified lambda value should be also analyze, because it may also heavily affect the simulation results. The differentiation of chaotic cases and limit cycles would be also important, and the improvement of some implementation details should be improved.

Acknowledgements This research was supported by the National Research, Development and Innovation Office (NKFIH) K108405 and by the EFOP-3.6.2-16-2017-00015 "HU-MATHS-IN; Intensification of the activity of the Hungarian Industrial Innovation Service Network" grant. Supported by the ÚNKP-17-4 New National Excellence Program of the Ministry of Human Capacities.

References

1. Papageorgiou, E.I. (ed.): Fuzzy Cognitive Maps for Applied Sciences and Engineering: From Fundamentals to Extensions and Learning Algorithms, vol. 54. Springer Science & Business Media (2013)
2. Miklós, F., Hatwágner, M.F., Kóczy, L.T.: Uncertainty tolerance and behavioral stability analysis of fixed structure fuzzy cognitive maps. In: Proceedings of 8th European Symposium on Computational Intelligence and Mathematics (ESCIM 2016), pp. 15–23. Sofia, Bulgaria (2016)
3. Axelrod, R. (ed.): Structure of Decision: The Cognitive Maps of Political Elites. Princeton University Press (1976)
4. Kosko, B.: Fuzzy cognitive maps. Int. J. Man Mach. Stud. **24**(1), 65–75 (1986)
5. Kosko, B.: Hidden patterns in combined and adaptive knowledge networks. Int. J. Approx. Reason. **2**(4), 377–393 (1988)
6. Zhang, W.R.: (Yin)(Yang) Bipolar fuzzy sets. In: The 1998 IEEE International Conference on Fuzzy Systems Proceedings and IEEE World Congress on Computational Intelligence, vol. 1, pp. 835–840. IEEE (1998)
7. Tsadiras, A.K.: Inference using binary, trivalent and sigmoid fuzzy cognitive maps. In: Proceedings of the 10th International Conference on Engineering Applications of Neural Networks (EANN 2007), vol. 284, pp. 327–334 (2007)
8. Parsopoulos, K.E., Papageorgiou, E.I., Groumpos, P.P., Vrahatis, M.N.: A first study of fuzzy cognitive maps learning using particle swarm optimization. In: The 2003 Congress on Evolutionary Computation, 2003 CEC'03, vol. 2, pp. 1440–1447. IEEE (2003)
9. Stylios, C.D., Groumpos, P.P.: Mathematical formulation of fuzzy cognitive maps. In: Proceedings of the 7th Mediterranean Conference on Control and Automation, pp. 2251–2261 (1999)
10. Lu, W., Pedrycz, W., Liu, X., Yang, J., Li, P.: The modeling of time series based on fuzzy information granules. Expert Syst. Appl. **41**(8), 3799–3808 (2014)
11. Nawa, N.E., Furuhashi, T.: A study on the effect of transfer of genes for the bacterial evolutionary algorithm. In: 1998 Second International Conference on Knowledge-Based Intelligent Electronic Systems, KES'98, vol. 3, pp. 585–590. IEEE (1998)
12. Nawa, N.E., Furuhashi, T.: Fuzzy system parameters discovery by bacterial evolutionary algorithm. IEEE Trans. Fuzzy Syst. **7**(5), 608–616 (1999)
13. Nawa, N.E., Hashiyama, T., Furuhashi, T., Uchikawa, Y.: A study on fuzzy rules discovery using pseudo-bacterial genetic algorithm with adaptive operator. In: IEEE International Conference on Evolutionary Computation, pp. 589–593. IEEE (1997)

On Experimental Efficiency
for Retraction Operator to Stem Basis

Gonzalo A. Aranda-Corral, Joaquín Borrego-Díaz, Juan Galán-Páez
and Alejandro Trujillo Caballero

Abstract In this paper, we introduce an implementation of an inference rule called "Independence Rule" which lets us reduce the size of knowledge basis based on the retraction problem. This implementation is made in a functional language, Scala, and specialized on attribute implications. We evaluate its efficiency related to the Stem Base generation.

1 Introduction

At the present days, big amount of data are being generated which are needed for automated tasks. These tasks can range from a single process to most complicated reasoning, from a computational point of view.

One way of approaching this problem could be reducing information before being included into reasoning tasks. These could be performed removing valid information as long as consistency were preserved. Other approach, and our proposal, could be building global reasoning engine and adapt it to every different situation.

From a more human point of view, a person has only one mind and is able to handle different scopes of information and making decisions according the situation it is. For example, a person acts differently if he is at home or in the office, and make different answers to the same questions, and all of them are valid in appropriate context.

To represent the information, we use (propositional) attribute implications obtained by means of Formal Concept Analysis (FCA) [5]. From a table of information and FCA algorithms, a set of attribute implications can be obtained, which is

G. A. Aranda-Corral (✉) · A. T. Caballero
Department of Information Technology, Universidad de Huelva, Huelva, Spain
e-mail: gonzalo.aranda@dti.uhu.es

J. Borrego-Díaz · J. Galán-Páez
Department of Computer Science and Artificial Intelligence, Universidad de Sevilla,
Sevilla, Spain

© Springer Nature Switzerland AG 2019
M. E. Cornejo et al. (eds.), *Trends in Mathematics and Computational Intelligence*, Studies in Computational Intelligence 796,
https://doi.org/10.1007/978-3-030-00485-9_8

called "Stem Basis". This basis represents the information contained in the original table and can be used for reasoning by means of Armstrong rules, for example.

Based on the retractor operator, also called "Independence Rule" [2], our aim is to build a contextual reasoner which could be computationally applied to these new situations, efficiently and preserving the logic consistency. Therefore, we develop a program in Scala, a functional language, implementing this operator and taking advantage of all possible properties in order to speed up the creation of the contextual reasoner faster than generating an equivalent reasoner from scratch.

2 Theoretical Foundations

2.1 Formal Concept Analysis (FCA)

It is not mandatory a full understanding of FCA [5], so we are going to do a brief introduction and to emphasize in attribute implications which are the connection to conservative retraction.

FCA is a mathematical tool for knowledge acquisition. It starts from the definition of data as a *formal context*, $M = (O, A, I)$, which consists of two sets, O (the *objects*) and A (the *attributes*) and a binary boolean relation $I \subseteq O \times A$. The main goal in FCA is the computation of the *concept lattice* and the attribute implications associated to the context.

An attribute implication is a pair of sets of attributes, written as $Y_1 \rightarrow Y_2$, where from a propositional logic point of view, $Y_1 \rightarrow Y_2$ is the formula $\bigwedge Y_1 \rightarrow \bigwedge Y_2$, and it is equivalent to a set of Horn clauses.

For every context, we can obtain a Stem basis [5], also called Duquenne–Guigues [6] base. This base is a set of implications which are complete and non-redundant set.

Actually one can choose $Y \rightarrow Y'' \setminus Y$ instead of $Y \rightarrow Y''$, so we will assume, by default, that for every implication $Y_1 \rightarrow Y_2$ which belongs to a stem basis, Y_1 and Y_2 are disjoint.

To illustrate this, we show an example of a formal context about living beings (see Fig. 1). In this example, we can see the objects (Cat, Leech, Frog, …), attributes (Need water, Aquatic, …) and a "×" when the relation holds between object and attribute. Applying the Duquenne–Guigues algorithm to calculate the stem basis, there are 6 attribute implications.

	Need water	Aquatic	Mobility	Legs	Animal	Plant
Cat	×		×	×	×	
Leech	×	×	×		×	
Frog	×	×	×	×	×	
Corn	×					×
Fish	×	×	×		×	
Bacteria	×	×	×		×	×

{ } ==> Need water;
Need water Aquatic ==> Mobility Animal;
Need water Mobility ==> Animal;
Need water Legs ==> Mobility Animal;
Need water Animal ==> Mobility;
Need water Mobility Animal Plant ==> Aquatic;

Fig. 1 Formal context and its attribute implications

2.2   Conservative Retractions

Conservative extensions is a popular problem that has been widely studied in Mathematical Logic. Based on this, we can reformulate the problem of conservative retractions analogously:

A theory T is a conservative extension of a theory T' (or T' is a conservative retraction) if every consequence of T in the language of T' is a consequence of T' already. This definition let us use smaller knowledge basis than original ones to perform contextual reasoning and preserve the logic consistency of results.

The main goal of this paper is to develop an efficient implementation of a retracting operator for computing a conservative retraction in a propositional logic theory. We will denote by $[T, L']$ a conservative retraction of T to the sublanguage L' throughout the paper.

We introduce an operator on propositional formulas as a translation of the usual derivation on $\mathbb{F}_2[\mathbf{x}]$ [2].

Definition 1 A $\partial : PForm \rightarrow PForm$ is a *Boolean derivation* if there exists a derivation d on the ring $\mathbb{F}_2[\mathbf{x}]$ such that map $\partial = \Theta \circ d \circ \pi$

If the derivation on $\mathbb{F}_2[\mathbf{x}]$ is $d = \frac{\partial}{\partial x_p}$, we denote ∂ as $\frac{\partial}{\partial p}$, we obtain:

$$\frac{\partial}{\partial p} F \equiv \neg(F\{p/\neg p\} \leftrightarrow F)$$

Thus, the value of $\frac{\partial}{\partial p} F$ with respect to a valuation does not depend on p, and the definition for the "independence rule" (or retraction operator) is:

Definition 2 The *independence rule* (or ∂-rule) on polynomial formulas is

$$\partial_x(a_1, a_2) : \quad \frac{a_1, \ a_2}{1 + \Phi\left[(1 + a_1 \cdot a_2)(1 + a_1 \cdot \frac{\partial}{\partial x}a_2 + a_2 \cdot \frac{\partial}{\partial x}a_1 + \frac{\partial}{\partial x}a_1 \cdot \frac{\partial}{\partial x}a_2)\right]}$$

For formulas the rule is translated as

$$\partial_p(F_1, F_2) := \Theta(\partial_{x_p}(\pi(F_1), \pi(F_2))).$$

2.3   Conservative Retraction on Attribute Implications

The Stem basis of a formal context is a set of attribute implications which is equivalent to it. Applying the conservative retraction operator to a stem basis, we can obtain a new set of attribute implications with a smaller language and logic consistency preserved.

In this way, we can obtain the stem basis for the full context, which is computationally hard, and adapt this basis to every contextual reasoning avoiding to recalculate a new basis from scratch. This can be justified as long as the calculation of the retraction is faster than the complete.

We obtain a definition for ∂_p for attribute implications by means of an extension of classic propositional resolution $res_p(\cdot, \cdot)$.

Lemma 1 *Let $C_i \equiv \bigwedge Y_1^i \rightarrow \bigwedge Y_2^i$ be an implication ($i = 1, 2$, $Y_1^i \cap Y_2^i = \emptyset$), and Γ be a set of implications. Let $\partial_p^c(C_1, C_2)$ be the symmetric operator*

$$\partial_p^c(C_1, C_2) := \begin{cases} \{C_1, C_2\} & p \notin var(C_1) \cup var(C_2) \\ \{C_2\} & p \in Y_1^1, \ p \notin var(C_2) \\ \{\bigwedge Y_1^1 \rightarrow \bigwedge(Y_2^1 \setminus \{p\}), C_2\} & p \in Y_2^1, \ p \notin var(C_2) \\ \{\top\} & p \in (Y_1^1 \cap Y_1^2) \cup (Y_2^1 \cap Y_2^2) \\ \{Resolvent_p(C_1, C_2)\} & p \in Y_1^1 \cap Y_2^2 \text{ or } p \in Y_1^2 \cap Y_2^1 \end{cases}$$

If $\partial_p^c[\Gamma] := \bigcup\{\partial_p^c(C_1, C_2) \ : \ C_1, C_2 \in \Gamma\}$, then $\partial_Q^c[\Gamma] \equiv \partial_Q[\Gamma] \, (Q \subseteq PV)$.

As we can see in Definition 1, when we apply this operator to all the formulas into the set, the variable used in retraction disappear from them. The language of the new set of formulas is a subset of the original.

3 Implementation

This implementation[1] of the retraction operator have been written in Scala, a functional and object-oriented language which let us to write source code near to mathematical definitions. Scala must be runned in a Java Virtual Machine, in order to be portable and integrable in most of platforms. For this, Scala empowers us to zip the code into a ".jar" file and redistribute it.

The software was developed from scratch. Firstly, we develop a small logical framework not restricted only in attribute implications. We develop a set of logic functions in order to check properties of our operator and validate the final implementation. For this reason, this framework could be extended to more logical functionality easily.

The original retraction operator is defined for 2 formulas (see Lemma 1) and is extended for a set of formulas in a natural way, taking the first advantage in its symmetry.

As we said before, the numbers of variables are decreasing after every applications but, on the other hand, the number of formulas increase. Note that in same cases of the operator it produces two formulas each step [4].

---

[1] http://protosmart.uhu.es/retraction/ImplicationRetractor.jar.

Optimizations This increase in formulas penalizes the perfomance of the system. Therefore, we focus all our efforts in this number.

After applying the operator, we analyze the results and mainly consider this optimizations:

- Implications with empty consequent. $Y_1 \to \top$
  This implication is a tautology and, therefore, it is always true. We can remove it from results.
- Implications with empty antecedent. $\top \to Y2$
  This formula implies that all attributes involved in Y_2 are True, triggering that we must replace the attributes with True. Because of our implementation, this replacing is equivalent to remove all appearances of the attributes into all formulas.
- Repeated formulas
  This was the hardest optimization because it induced us to develop extra data structures and increase our memory consumption in order to preserve the execution time.

These optimizations, and some minors, do not reduce the complexity of this problem, but are aimed to reduce the time of computation. Memory is not a big deal but time is.

4 Experiments and Results

In order to analyze the proper operation of the algorithm as well as the optimizations, both versions have been executed on different sets of rules. These sets of rules were generated from random FCA contexts and calculating its stem basis. Random contexts were also experimented in different sizes.

First experiment was designed to investigate how the performance of the algorithm retracting all variables in the language is. We created several squared random FCA contexts with 7 attributes and a density of relation of 0.4. After generating all set of rules, we counted the time it took for every retraction and results are shown in Fig. 2.

We note that for first iterations, when the number of formulas did not grow too, original algorithm has a little better performance than the optimized. The cost of optimizations are not advised for systems with an small set of formulas. After that, we can see that optimizations are a key point in the performance of the system.

Second experiment was devoted to check if it would be suitable for contextual reasoning. For this, we realized and experiment, in one hand, generating stem basis for FCA random context with n attributes and, on the other hand, starting from context with $n + 1$ attributes and retracting in 1 attribute.

As we see in Fig. 3, from a certain number of attributes the retractor operator is much faster than the generation from scratch. The improvement of optimizations are up to 65% with this number of attributes. Also we can see that the exponential increment of time is bigger in gencration more than retraction.

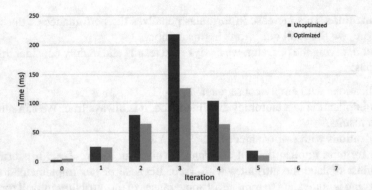

Fig. 2 Time consumed per iteration

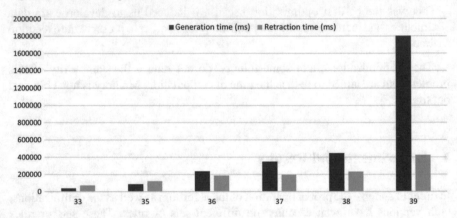

Fig. 3 Generation and retraction times

5 Conclusions and Future Works

In this paper, we introduce an efficient implementation of a retraction operator. The aim of this work is building mechanisms to automatize the contextual reasoning adapting the big knowledge basis into affordable problems. We also present several experiments which show that this way looks like a good starting point to keep working on that.

Our next step will be to extend this method for clauses and try to offer an efficient solution for SAT problem, introducing negation. Other future work could be implement this system in a "Map-Reduce" architecture. Based on the "independence" of operator's application to pairs of equations, it would be suitable for a big parallelization of the problem.

References

1. Alonso-Jiménez, J.A., Aranda-Corral, G.A., Borrego-Díaz, J., Fernández-Lebrón, M.M., Hidalgo-Doblado, M.J.: Extending attribute exploration by means of boolean derivatives. In: Proceedings of 6th International Conference on Concept Lattices and Their Applications (CLA2008), pp. 121–132 (2008)
2. Aranda-Corral, G.A., Borrego-Díaz, J., Fernández-Lebrón, M.M.: Conservative retractions of propositional logic theories by means of boolean derivatives: theoretical foundations. In Proceedings of 16th CALCULEMUS Congress. Lecture Notes in Artificial Intelligence, vol. 5625, pp. 45–58. Springer, Berlin (2009)
3. Borrego-Díaz, J., Fernández-Lebrón, M.: Theoretical Foundations of a Specialised Polynomial-Based Calculus for Computing Conservative Retractions in Propositional Logic (2008)
4. Cook, S.A.: The complexity of theorem-proving procedures. In: Proceedings of the 3rd Annual ACM Symposium on Theory of Computing, STOC 1971, pp. 151–158. ACM, New York, USA (1971)
5. Ganter, B., Wille, R.: Formal Concepts Analysis. Mathematical Foundations. Springer, Berlin (1999)
6. Guigues, J.-L., Duquenne, V.: Familles minimales d' implications informatives resultant d'un tableau de donnees binaires. Math. Sci. Hum. **95**, 5–18 (1986)
7. Papadimitriou, C.: Computational Complexity. Addison-Wesley, New York (1994)
8. Taouil, R., Bastide, Y.: Computing proper implications. In: Mephu et al., E. (eds.) Proceedings of Workshop on Concept Lattices-based Theory, Methods and Tools for Knowledge Discovery in Databases, pp. 49–61 (2001)
9. Yevtushenko, S.A.: System of data analysis "Concept Explorer". In: Proceedings of the 7th National Conference on Artificial Intelligence KI–2000, pp. 127–134 (2000)

Inference of Mixed Information in Formal Concept Analysis

P. Cordero, M. Enciso, A. Mora and J. M. Rodriguez-Jiménez

Abstract Negative information can be considered twofold: by means of a negation operator or by capturing the absence of information. In this second approach, a new framework have to be developed: from the syntax to the semantics, including the management of such generalized knowledge representation. In this work we traverse all these issues in the framework of formal concept analysis, introducing a new set of inference rules to manage mixed (positive and negative) attributes.

1 Introduction

We focus on Formal Concept Analysis which represents the information as a binary relation (named formal context) between objects as rows and attributes as columns. From this table, mining techniques to extract knowledge are well know and it is possible to compute a concept lattice and also a sets of implications representing the same knowledge. We are particularly interested in implications because they allow a symbolic manipulation by using logic.

Normally, the formal context stores that "an object has an attribute". In this work, we consider interesting not only this observation, represented with positive attributes, but also that "an object has not an attribute" (negative attribute), that is, not only the presence but also the absence of a given property.

This idea already appeared in data mining area, where some works [4] considered the negation of attributes inside of implications: the use of positive and negative

P. Cordero · M. Enciso · A. Mora (✉) · J. M. Rodriguez-Jiménez
University of Málaga, Andalucía Tech, Málaga, Spain
e-mail: amora@ctima.uma.es

P. Cordero
e-mail: pcordero@uma.es

M. Enciso
e-mail: enciso@uma.es

J. M. Rodriguez-Jiménez
e-mail: jmrodriguez@ctima.uma.es

© Springer Nature Switzerland AG 2019   81
M. E. Cornejo et al. (eds.), *Trends in Mathematics and Computational
Intelligence*, Studies in Computational Intelligence 796,
https://doi.org/10.1007/978-3-030-00485-9_9

attributes appeared in basket market analysis, when rules considering that "if a customer usually buys a given product then he does not buy other specific product". In these former works, the negative information negatively impacts on the data mining method because some cut mechanisms used became not valid. In our opinion, this situation can be resolved if a suitable treatment with positive and negative attributes are added.

In FCA, as far as we know, only some works consider the management of both positive and negative attributes in FCA. In [3], the authors join the formal context with its complementary and use the classical FCA methods of data mining to retrieve implications with positive and negative attributes. Unfortunately, such an approach cannot take advantage of the semantics of the enriched framework. In 2012, Missaoui et al. [5] have approached the generation of mixed implications from two given sets of implications only with positive and negative attributes respectively. We have followed this line: in [11], we extended the classical FCA framework with new derivation operators constituting a Galois connection for the treatment of negative and positive information in FCA and in [9, 10] we proposed some mining algorithms to derive directly mixed implications.

In this work, we progress in this line by proposing a set of inference rules to manage efficiently mixed implications. This logic is based on Simplification paradigm presented in [6].

The structure of the paper is the following: Sect. 2 shows some preliminaries. The new set of rules to manipulate implications with positive and negative attributes is presented in Sect. 3. Finally, some conclusions appear in Sect. 4.

2 Preliminaries

The data in Formal Concept Analysis are stored in a binary table named a *formal context* which formally is defined as the triple $\mathbb{K} = \langle G, M, I \rangle$ where G and M are finite non-empty sets and $I \subseteq G \times M$ is a binary relation. The elements in G are named objects, the elements in M attributes and $\langle g, m \rangle \in I$ means that the object g has the attribute m.

In this triple, two mappings $\uparrow : 2^G \to 2^M$ and $\downarrow : 2^M \to 2^G$, named derivation operators, are defined as follows: for any $X \subseteq G$ and $Y \subseteq M$,

$$X^\uparrow = \{m \in M \mid \langle g, m \rangle \in I \text{ for all } g \in X\} \tag{1}$$

$$Y^\downarrow = \{g \in G \mid \langle g, m \rangle \in I \text{ for all } m \in Y\} \tag{2}$$

$X^\uparrow$ is the subset of all attributes shared by all the objects in X and $Y^\downarrow$ is the subset of all objects that have the attributes in Y. The pair $(\uparrow, \downarrow)$ constitutes a Galois connection between 2^G and 2^M and, therefore, both compositions are closure operators.

A pair of subsets $\langle X, Y \rangle$ with $X \subseteq G$ and $Y \subseteq M$ such that $X^\uparrow = Y$ and $Y^\downarrow = X$ is named a *formal concept* where X is its *extent* and Y its *intent*. These extents and intents coincide with closed sets wrt the closure operators because $X^{\uparrow\downarrow} = X$ and $Y^{\downarrow\uparrow} = Y$. Thus, the set of all formal concepts is a lattice, named *concept lattice*, with the relation

$$\langle X_1, Y_1 \rangle \leq \langle X_2, Y_2 \rangle \text{ if and only if } X_1 \subseteq X_2 \text{ (or equivalently, } Y_2 \subseteq Y_1) \quad (3)$$

This concept lattice will be denoted by $\mathfrak{B}(\mathbb{K})$. The concept lattice can be characterized in terms of *attribute implications* being expressions $A \rightarrow B$ where $A, B \subseteq M$. An implication $A \rightarrow B$ holds in a context $\mathbb{K}$ if $A^{\downarrow} \subseteq B^{\downarrow}$. That is, any object that has all the attributes in A has also all the attributes in B. These implications can be syntactically managed in a logical style [1]. The main aim of this paper is to provide a sound and complete logic to manage implications that also considers negative information.

3 Simplification Paradigm and Mixed Attributes

Armstrong's axioms [1] constitutes a pioneer complete axiomatic system which opened the door to the manipulation of implications. It was designed to characterize the semantics of implications but it was not conceived to design automated method around it. In fact, its main pillar is the transitivity paradigm, which avoids the design of efficient automated methods. In [2] the authors presents, for first time, an alternative paradigm to tackle this issue, by introducing the Simplification paradigm and building a new complete axiomatic system named Simplification Logic for Functional Dependencies $\mathbf{SL}_{FD}$. It avoids the use of transitivity and is guided by the idea of simplifying the set of implications by efficiently removing some redundant attributes [2].

We define the $\mathbf{SL}_{FD}$ logic as the pair $(\mathcal{L}_S, \mathcal{S}_S)$, corresponding to its language and axiomatic system.

Definition 1 Let M be a set of attributes, the set of well formed formulas of $\mathbf{SL}_{FD}$ is defined as $\mathcal{L}_S = \{X \rightarrow Y \mid X, Y \subseteq M\}$.

Once the language has been defined, semantics provides an interpretation for each well formed formula of the language. Semantics of implications can be introduced in several frameworks. In this work we consider Formal Concept Analysis to introduce $\mathbf{SL}_{FD}$ semantics.

Definition 2 Let $\mathbb{K} = \langle G, M, I \rangle$ be a formal context and $A \rightarrow B \in \mathcal{L}_S$. The context $\mathbb{K}$ is said to be a *model* for $A \rightarrow B$ if $B \subseteq A^{\downarrow\uparrow}$. It is denoted by $\mathbb{K} \models A \rightarrow B$.

We remark that

$$\mathbb{K} \models A \rightarrow B \quad \text{if and only if} \quad A^{\downarrow} \subseteq B^{\downarrow}$$

As usual, the notion of models can be extended to implicational systems: given $\Sigma \subseteq \mathcal{L}_S$, the expression $\mathbb{K} \models \Sigma$ means that $\mathbb{K} \models A \to B$ for all $A \to B \in \Sigma$.

Definition 3 Let M be a set of attributes, $A \to B \in \mathcal{L}_S$ and $\Sigma \subseteq \mathcal{L}_S$. The implication $A \to B$ is said to be *semantically derived* from Σ, denoted by $\Sigma \models A \to B$, if $\mathbb{K} \models \Sigma$ implies $\mathbb{K} \models A \to B$ for every formal context $\mathbb{K}$.

On the other hand, two implicational systems $\Sigma_1, \Sigma_2 \subseteq \mathcal{L}_S$ are *semantically equivalent*, denoted by $\Sigma_1 \equiv \Sigma_2$, if the following equivalence holds

$$\mathbb{K} \models \Sigma_1 \text{ if and only if } \mathbb{K} \models \Sigma_2$$

for every formal context $\mathbb{K}$.

In summary, $\Sigma \models A \to B$ if every model for Σ is a model for $A \to B$ and $\Sigma_1 \equiv \Sigma_2$ if their models are the same, it means, both sets represent the same knowledge. Regarding the axiomatic system, $\mathcal{S}_S$ considers reflexivity as axiom scheme

$$[\text{Ref}] \ \frac{A \supseteq B}{A \to B}$$

together with the following inference rules called fragmentation, composition and simplification respectively.

$$[\text{Frag}] \ \frac{A \to B \cup C}{A \to B} \quad [\text{Comp}] \ \frac{A \to B, \ C \to D}{A \cup C \to B \cup D} \quad [\text{Simp}] \ \frac{A \to B, \ C \to D}{A \cup (C \smallsetminus B) \to D}$$

The equivalence between $\mathbf{SL}_{FD}$ logic and Armstrong's Axioms and an efficient algorithm to compute the closure of a set of attributes were proposed in [6].

In this section, we introduce a natural extension of Simplification logic to consider positive and negative information. As we have done in the classical case, we introduce the language and a sound and complete axiomatic system.

3.1 The Language of SL_{Mx}

Lowercase character m will be used to denote positive attributes and $\overline{m}$ will denote the negation of the attribute m. $\overline{M}$ denotes the set $\{\overline{m} \mid m \in M\}$ whose elements will be named negative attributes.

Definition 4 Given a finite set of attributes M, the language of SL_{Mx} is

$$\mathcal{L}_S = \{X \to Y \mid X, Y \subseteq M \cup \overline{M}\}.$$

Formulas in $\mathcal{L}_S$ are named *mixed attribute implications*.

Arbitrary elements in $M \cup \overline{M}$ are denoted by using the first letters in the alphabet: a, b, c, etc. For each $a \in M \cup \overline{M}$, $\overline{a}$ denotes its opposite. Namely, the symbol a could represent a positive or a negative attribute i.e. if $a = m \in M$ then $\overline{a} = \overline{m}$ and if $a = \overline{m} \in \overline{M}$ then $\overline{a} = m$.

Similarly as we did in the classical framework, the subsets of $M \cup \overline{M}$ will be denoted by uppercase characters A, B, C, etc. and we introduce the following notation: for each $A \subseteq M \cup \overline{M}$,

- $\overline{A}$ is the set of the opposite of attributes in A, i.e. $\{\overline{a} \mid a \in A\}$
- $\mathrm{Pos}(A) = \{m \in M \mid m \in A\}$ and $\mathrm{Neg}(A) = \{m \in M \mid \overline{m} \in A\}$
- $\mathrm{Tot}(A) = \mathrm{Pos}(A) \cup \mathrm{Neg}(A)$

Note that $\mathrm{Pos}(A), \mathrm{Neg}(A), \mathrm{Tot}(A) \subseteq M$.

3.2 The Semantics of SL_{Mx}

Now we introduce the semantics of SL_{Mx}. Classical derivation operators are extended as follows:

Definition 5 Let $\mathbb{K} = \langle G, M, I \rangle$ be a formal context. We define the operators $\Uparrow$: $2^G \to 2^{M \cup \overline{M}}$ and $\Downarrow : 2^{M \cup \overline{M}} \to 2^G$ as follows: for $X \subseteq G$ and $Y \subseteq M \cup \overline{M}$,

$$X^{\Uparrow} = \{m \in M \mid \langle g, m \rangle \in I \text{ for all } g \in X\} \cup \{\overline{m} \in \overline{M} \mid \langle g, m \rangle \notin I \text{ for all } g \in X\}$$
$$Y^{\Downarrow} = \{g \in G \mid \langle g, m \rangle \in I \text{ for all } m \in Y\} \cap \{g \in G \mid \langle g, m \rangle \notin I \text{ for all } \overline{m} \in Y\}$$

The above extended derivation operators have similar properties as the classical ones. More specifically, they define a Galois connection:

Theorem 1 *For any formal context* $\mathbb{K} = \langle G, M, I \rangle$, *the pair* $(\Uparrow, \Downarrow)$ *is a Galois connection between* $(2^G, \subseteq)$ *and* $(2^{M \cup \overline{M}}, \subseteq)$.

As a consequence of the above theorem, similarly that occurs in the classical case, both compositions $\Uparrow \circ \Downarrow$ and $\Downarrow \circ \Uparrow$ are closure operators and lead to notions of mixed formal concept and mixed concept lattice [9].

Now, we can provide a meaning for formulas in the language (implications).

Definition 6 Let $\mathbb{K} = \langle G, M, I \rangle$ be a formal context and $A \to B \in \mathcal{L}_S$. The context $\mathbb{K}$ is a model for $A \to B$, denoted by $\mathbb{K} \models A \to B$, if $A^{\Downarrow} \subseteq B^{\Downarrow}$, or equivalently $B \subseteq A^{\Downarrow\Uparrow}$.

Example 1 Considering the formal context $\mathbb{K} = \langle G, M, I \rangle$ where the set of objects is $G = \{o_1, o_2, o_3, o_4\}$, the set of attributes is $M = \{m_1, m_2, m_3, m_4, m_5\}$ and I is the binary relation depicted in Table 1, we have that $\mathbb{K} \not\models m_2 \to m_4$ and $\mathbb{K} \models m_2 \to \overline{m}_4$ whereas $\mathbb{K} \not\models m_2 \to m_3$ either $\mathbb{K} \not\models m_2 \to \overline{m}_3$.

As usual, given a set of mixed attribute implications $\Sigma \subseteq \mathcal{L}_S$ and a formal context $\mathbb{K}$, the expression $\mathbb{K} \models \Sigma$ denotes $\mathbb{K} \models A \to B$ for all $A \to B \in \Sigma$ and $\Sigma \models A \to B$ denotes that any model for Σ is also model for $A \to B$.

Table 1 A formal context

I	m_1	m_2	m_3	m_4	m_5
o_1		×	×		×
o_2	×	×			
o_3		×	×		×
o_4			×	×	

3.3   An Axiomatic System for SL_{Mx}

To end this section, we propose a set of inference rules to reason with mixed implications strongly inspired by the simplification paradigm.

The axiomatic system for SL_{Mx} considers two axiom schemes and four inference rules. They are the following:

[Ref] Reflexivity: $\vdash_{Mx} A \to A$.
[Simp] Simplification: $A \to B, C \to D \vdash_{Mx} A(C - B) \to D$.
[Key] Key: $A \to b \vdash_{Mx} A\overline{b} \to M\overline{M}$.
[RevKey] Reverse Key: $Ab \to M\overline{M} \vdash_{Mx} A \to \overline{b}$.
[Red] Reduction: $Ab \to C, A\overline{b} \to C \vdash_{Mx} A \to C$.

4   Conclusions and Future Works

We propose a new logic for the automated and efficient treatment of implications considering positive and negative information. It opens the door for the future developments of methods in FCA and in other areas.

Some interesting future research lines regarding a further generalization of our proposal, could be addressed in The Three-Way Formal Concept Analysis [7, 8] exploits the idea of dividing the universe of discourse into three disjoint subsets: positive, negative and boundary. The idea of using negative information in the contexts is approached in these works by means of positive and negative operators.

Acknowledgements Supported by project regional no. TIN2014-59471-P of the Science and Innovation Ministry of Spain, co-funded by the European Regional Development Fund (ERDF).

References

1. Armstrong, W.: Dependency structures of data base relationships. In: Proceedings of IFIP Congress, pp. 580–583. North Holland, Amsterdam (1974)
2. Cordero, P., Enciso, M., Mora, A.: de Guzmn, I.P.: Slfd logic: elimination of data redundancy in knowledge representation. In: IBERAMIA. Lecture Notes in Computer Science, vol. 2527, pp. 141–150. Springer, Berlin (2002)

3. Gasmi, G., Yahia, S.B., Nguifo, E.M., Bouker, S.: Extraction of association rules based on literalsets. In: LNCS, vol. 4654, pp. 293–302 (2007)
4. Mannila, H., Toivonen, H., Verkamo, A.I.: Efficient algorithms for discovering association rules. In: KDD Workshop. pp. 181–192 (1994)
5. Missaoui, R., Nourine, L., Renaud, Y.: Computing implications with negation from a formal context. Fundam. Inf. **115**(4), 357–375 (2012)
6. Mora, A., Enciso, M., Cordero, P.: Closure via functional dependence simplification. Int. J. Comput. Math. **89**, 510–526 (2012)
7. Qi, J., Qian, T., Wei, L.: The connections between three-way and classical concept lattices. Knowl Based Syst. **91**, 143–151 (2016)
8. Ren, R., Wei, L.: The attribute reductions of three-way concept lattices. Knowl Based Syst. **99**, 92–102 (2016)
9. Rodríguez-Jiménez, J., Cordero, P., Enciso, M., Mora, A.: A generalized framework to consider positive and negative attributes in formal concept analysis. In: CLA. CEUR Workshop Proceedings, vol. 1252, pp. 267–278 (CEUR-WS.org) (2014)
10. Rodríguez-Jiménez, J., Cordero, P., Enciso, M., Mora, A.: Data mining algorithms to compute mixed concepts with negative attributes: an application to breast cancer data analysis. Math. Methods Appl. Sci. (2016)
11. Rodriguez-Jimenez, J., Cordero, P., Enciso, M., Rudolph, S.: Concept lattices with negative information: a characterization theorem. Inf. Sci. (2016). https://doi.org/10.1016/j.ins.2016.06.015

Unifying Reducts in Formal Concept Analysis and Rough Set Theory

M. José Benítez-Caballero, Jesús Medina and Eloísa Ramírez-Poussa

Abstract Attribute reduction is a fundamental part in different mathematical tools devoted to data analysis, such as, Rough Set Theory and Formal Concept Analysis. These last mathematical theories are closely related and, in this paper, we establish connections between attribute reduction in both frameworks. Mainly, we have introduced a sufficient and necessary condition in order to ensure that the reducts in both theories coincide.

1 Introduction

Rough Set Theory (RST) and Formal Concept Analysis (FCA) are two powerful mathematical tools for processing incomplete information in large databases, which contain an objects set, an attributes set and a relationship between them.

Reduce the set of attributes, keeping the obtained information of the original database, is one of the most important goals in both frameworks. However, attribute reduction considers a different philosophy in both theories. In FCA we reduce the number of attributes of the database without modifying the structure of the original concept lattice, while in RST we select the attributes without losing the ability to discern objects, this last procedure is usually called attribute selection. Hence, the philosophies of both reductions are not equivalent. Although the attribute reduction procedures in FCA and in RST are not equivalent. Despite this lack of similarity, study

Partially supported by the State Research Agency (AEI) and the European Regional Development Fund (ERDF) project TIN2016-76653-P.

M. J. Benítez-Caballero · J. Medina · E. Ramírez-Poussa (✉)
Department of Mathematics, Universidad de Cádiz, Cádiz, Spain
e-mail: eloisa.ramirez@uca.es

M. J. Benítez-Caballero
e-mail: mariajose.benitez@uca.es

J. Medina
e-mail: jesus.medina@uca.es

© Springer Nature Switzerland AG 2019
M. E. Cornejo et al. (eds.), *Trends in Mathematics and Computational Intelligence*, Studies in Computational Intelligence 796,
https://doi.org/10.1007/978-3-030-00485-9_10

mathematical results relating both procedures is interesting, since the contributions in one framework enrich the other one.

Attribute reduction has been studied separately in both theories [4, 5, 7, 9, 11, 12], although only a few papers study the existing connections [3, 15], showing that the reducts in both frameworks do not coincide, in general.

The main contribution of this paper is to introduce a sufficient and necessary condition in order to ensure that the reducts in RST and in FCA, obtained from the considered database, are the same.

2 Preliminaries

Let us begin recalling the necessary notions and results of RST and FCA. First of all, we need to recall the notion of information system.

Definition 1 An *information system* $(U, \mathcal{A})$ is a tuple, such that the sets $U = \{x_1, x_2, \ldots, x_n\}$ and $\mathcal{A} = \{a_1, a_2, \ldots, a_m\}$ are finite, non-empty sets of objects and attributes, respectively, in which, each $a \in \mathcal{A}$ corresponds to a mapping $\bar{a} : U \to V_a$, where V_a is the value set of a over U. For every subset D of $\mathcal{A}$, the D-*indiscernibility relation*, $\text{Ind}(D)$, is defined as the equivalence relation $\text{Ind}(D) = \{(x_i, x_j) \in U \times U \mid \text{for all } a \in D, \bar{a}(x_i) = \bar{a}(x_j)\}$.

If we have that the value set of a is $V_a = \{0, 1\}$, for all $a \in \mathcal{A}$, $(U, \mathcal{A})$ is called *boolean information system*.

The notions of consistent set and reduct of an information system are needed to relate RST and FCA.

Definition 2 Let $(U, \mathcal{A})$ be an information system and a subset of attributes $D \subseteq \mathcal{A}$. D is a *consistent set* of $(U, \mathcal{A})$ if $\text{Ind}(D) = \text{Ind}(\mathcal{A})$. Moreover, if for each $a \in D$ we have that $\text{Ind}(D \setminus \{a\}) \neq \text{Ind}(\mathcal{A})$, then D is called *reduct* of $(U, \mathcal{A})$.

Definition 3 Given an information system $(U, \mathcal{A})$, its *discernibility matrix* is a matrix of order $|U| \times |U|$, denoted as $M_{\mathcal{A}}$, in which the element $M_{\mathcal{A}}(i, j)$ for each pair of objects (i, j) is defined by $M_{\mathcal{A}}(i, j) = \{a \in \mathcal{A} \mid \bar{a}(i) \neq \bar{a}(j)\}$ and the *discernibility function* of $(U, \mathcal{A})$ is defined by:

$$\tau_{\mathcal{A}} = \bigwedge \left\{ \bigvee (M_{\mathcal{A}}(i, j)) \mid i, j \in U \text{ and } M_{\mathcal{A}}(i, j) \neq \varnothing \right\}$$

Reducts in RST are characterized from the notion of discernibility function, as the following result shows.

Theorem 1 *Given a boolean information system* $(U, \mathcal{A})$. *An arbitrary set* D, *where* $D \subseteq \mathcal{A}$, *is a reduct of the information system if and only if the cube* $\bigwedge_{a \in D} a$ *is a cube in the restricted disjunctive normal form*[1] *(RDNF) of the discernibility function* $\tau_{\mathcal{A}}$.

---

[1]We assume that the reader is familiar with the notions related to classical theory of propositional logic [6, 8].

Now, the basic definitions of FCA, will be recalled. In FCA, a *context* is a triple (A, B, R) composed of the attributes set A, the objects set B and the relationship $R: A \times B \rightarrow \{0, 1\}$, defined, for each $a \in A$ and $b \in B$, as $R(a, b) = 1$, if a and b are related and $R(a, b) = 0$, otherwise. From a context, the *concept-forming operators* are the mappings $\uparrow: 2^B \rightarrow 2^A$, $\downarrow: 2^A \rightarrow 2^B$ defined for each $X \subseteq B$ and $Y \subseteq A$, as:

$$X^\uparrow = \{a \in A \mid \text{for all } b \in X, a R b\} = \{a \in A \mid \text{if } b \in X, \text{ then } a R b\} \quad (1)$$
$$Y^\downarrow = \{b \in B \mid \text{for all } a \in Y, a R b\} = \{b \in B \mid \text{if } a \in Y, \text{ then } a R b\} \quad (2)$$

A *concept* is a pair (X, Y), where $X \subseteq B$, $Y \subseteq A$, satisfying that $X^\uparrow = Y$ and $Y^\downarrow = X$. The *extent* is the subset of objects X of the concept (X, Y) and the subset of attributes Y is called *intent*. The set of all the concepts, denoted as $\mathcal{B}(A, B, R)$, with the inclusion ordering on the left argument, is a complete lattice [6, 9].

Due to the mappings $\uparrow$ and $\downarrow$ form a Galois connection, given an object $b \in B$, we can define an *object-concept* as the concept generated by b, that is, $(b^{\uparrow\downarrow}, b^\uparrow)$. Analogously, an *attribute-concept*, $(a^\downarrow, a^{\downarrow\uparrow})$, is the concept generated by an attribute $a \in A$.

On the other hand, the main goal of attribute reduction in FCA is to reduce the set of attributes without changing the structure of the concept lattice.

Given a context (A, B, R) and a subset of attributes $Y \subseteq A$, the triple $(Y, B, R_{|Y})$ is also a formal context, where $R_{|Y} = R \cap (Y \times B)$ denotes the restricted relation. Then, we can define the restricted mappings $\downarrow^Y$ and $\uparrow^Y$, in a similar way to the ones given in Eqs. (1) and (2). It is clear that $X^{\uparrow_Y} = X^\uparrow \cap Y$, for any subset of objects $X \subseteq B$. Finally, we recall the notions of consistent set and reduct in the FCA framework.

Definition 4 Let (A, B, R) be a context, if there exists a set of attributes $Y \subseteq A$ such that $\mathcal{B}(A, B, R) \cong \mathcal{B}(Y, B, R_{|Y})$, then Y is called a *consistent* set of (A, B, R). Moreover, if $\mathcal{B}(Y \smallsetminus \{y\}, B, R_{|Y \smallsetminus \{y\}}) \ncong \mathcal{B}(A, B, R)$, for all $y \in Y$, then Y is called *reduct* of (A, B, R).

3 Relating Reducts in RST and FCA

In this section, we will consider a finite set of attributes and of objects. In addition, to clarify the environment in which we are working, RST or FCA, we will denote a reduct of an information system $(U, \mathcal{A})$ as *RS-reduct* and a reduct of the context (A, B, R) as *CL-reduct*. Similarly, a consistent set of the information system $(U, \mathcal{A})$ will be called as *RS-consistent set* and a consistent set of the context (A, B, R) as *CL-consistent set*.

First of all, we will recall how to obtain an information system from a formal context and a technical lemma, which are needed to relate the considered operators.

Definition 5 Let (A, B, R) be a context, a *context information system* is defined as the pair (B, A) where the mappings $\bar{a} : B \to V_a$, with $V_a = \{0, 1\}$, are defined as $\bar{a}(b) = R(a, b)$, for all $a \in A, b \in B$.

Lemma 1 ([3]) *Given a context (A, B, R) and the corresponding context informa-tion system (B, A), the equality $a^\downarrow = \bar{a}$ holds, for each $a \in A$.*

In addition, in [15] it was proven that a CL-consistent set of a context always provides an RS-consistent set of the associated context information system.

Theorem 2 ([15]) *Given a context (A, B, R) and the corresponding context infor-mation system (B, A). If $D \subseteq A$ is a CL-consistent set then D is an RS-consistent set.*

The counterpart of Theorem 2 is not true, in general, as we can see in the following example. More properties can be seen in [3, 15]. Therefore, both notions are not equivalent, even more so when we consider RS-reducts and CL-reducts.

Example 1 We will consider a context (A, B, R) composed of two objects $b_1, b_2 \in B$, two attributes $a_1, a_2 \in A$, and the relationship R defined in Fig. 1.

In this case, we have that $a_1^\downarrow = \{b_1\}$ and $a_2^\downarrow = \{b_2\}$. Moreover, we can see in Fig. 1 that $C_1 = (a_1^\downarrow, a_1^{\downarrow\uparrow})$ and $C_2 = (a_2^\downarrow, a_2^{\downarrow\uparrow})$ are meet-irreducible elements. Hence, we cannot remove these concepts of the concept lattice and the attributes a_1 and a_2 are necessary to preserve the structure of the concept lattice. Thus, the CL-reduct of this context is the set $\{a_1, a_2\}$.

If we consider the context information system corresponding to the context (A, B, R), we obtain that the following discernibility matrix:

$$\begin{pmatrix} \varnothing & \{a_1, a_2\} \\ \{a_1, a_2\} & \varnothing \end{pmatrix}$$

It is easy to see that, in this case, the discernibility function is given by $\tau_A = \{a_1\} \vee \{a_2\}$. Therefore, if we want to discern the objects b_1 and b_2 we need to consider one attribute of the context information system. In this way, two different RS-reducts $D_1 = \{a_1\}$ and $D_2 = \{a_2\}$ $(\text{Ind}(D_1) = \text{Ind}(D_2) = \text{Ind}(A))$ are obtained, whilst in FCA we have to take into account both attributes.   □

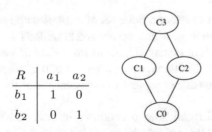

R	a_1	a_2
b_1	1	0
b_2	0	1

Fig. 1 Relation R and the Hasse diagram of the context in Example 1

Now, we are interested in establishing the constraints to ensure that the family of CL-reducts of a context coincides with the family of RS-reducts of the associated context information system. For that purpose, we need to consider an special kind of formal context. Specifically, we say that *a context* (A, B, R) *does not have cross-values* if for every $a_1, a_2 \in A, i, j \in B$ satisfying $R(a_1, i) \neq R(a_1, j)$ and $R(a_2, i) \neq R(a_2, j)$, then the equalities $R(a_1, i) = R(a_2, i)$ and $R(a_1, j) = R(a_2, j)$ holds. Every context with no cross-values satisfies the following property:

Proposition 1 *If the context* (A, B, R) *does not have cross-values, then all attribute-concepts in the associated concept lattice are comparable.*

The following result fixes the condition to guarantee that a CL-reduct of a context is a RS-reduct of the associated context information system.

Theorem 3 *Let* (A, B, R) *be a context and* (B, A) *be the corresponding context information system. The family of CL-reducts coincides with the family of RS-reducts if and only if the context* (A, B, R) *does not have cross-values.*

The following example illustrates the previous result.

Example 2 Let (A, B, R) be a context with three objects $b_1, b_2, b_3 \in B$, four attributes $a_1, a_2, a_3, a_4 \in A$, and the relationship R displayed in Fig. 2.

The associated concept lattice can be seen on the right side of Fig. 2, where we can observe that the set of meet-irreducible elements is the following:

$$C_0 = (a_2^{\downarrow}, a_2^{\downarrow\uparrow}) = (a_4^{\downarrow}, a_4^{\downarrow\uparrow}) = (\{b_1\}, \{a_1, a_2, a_3, a_4\})$$
$$C_1 = (a_1^{\downarrow}, a_1^{\downarrow\uparrow}) = (\{b_1, b_3\}, \{a_1, a_3\})$$

Therefore, we cannot remove these concepts of the context and, in order to preserve the original structure of the concept lattice, we need to consider the subset of attributes $D_1 = \{a_1, a_2\}$ or $D_2 = \{a_1, a_4\}$. Thus, the sets D_1 and D_2 are the CL-reducts of this context.

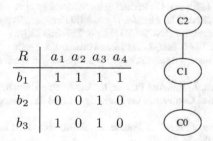

R	a_1	a_2	a_3	a_4
b_1	1	1	1	1
b_2	0	0	1	0
b_3	1	0	1	0

Fig. 2 Relation R and the Hasse diagram of the context in Example 2

Considering the context information system associated with the context (A, B, R), we obtain that the discernibility matrix is:

$$\begin{pmatrix} \varnothing & \{a_1, a_3, a_4\} & \{a_2, a_4\} \\ \{a_1, a_3, a_4\} & \varnothing & \{a_1\} \\ \{a_2, a_4\} & \{a_1\} & \varnothing \end{pmatrix}$$

From this discernibility matrix, the obtained discernibility function is $\tau_A = \{a_1, a_2\} \vee \{a_1, a_4\}$. Hence, according to Theorem 1 the RS-reducts are $D'_1 = \{a_1, a_2\}$ and $D'_2 = \{a_1, a_4\}$, which coincide with the CL-reducts. Therefore, in this case, both families coincide.   □

4   Conclusions and Future Work

In this paper, we have progressed in the study of the relationship between attribute reduction in FCA and RST. Mainly, we have presented an interesting result which set a sufficient and necessary conditions to ensure that the family of RS-reducts and the family of CL-reducts are the same. We have also introduced detailed examples in order to illustrate the contributions.

Bireducts were presented as a natural extension of reducts in RST. Bireducts reduce attributes and objects at the same time, providing more flexibility [1, 2, 10, 13, 14]. In the future, we will extend this notion to the FCA framework.

References

1. Benítez, M., Medina, J., Ślęzak, D.: Delta-information reducts and bireducts. In: Alonso, J.M., Bustince, H., Reformat, M. (eds.) 2015 Conference of the International Fuzzy Systems Association and the European Society for Fuzzy Logic and Technology (IFSA- EUSFLAT-15), Gijón, Spain, pp. 1154–1160. Atlantis Press (2015)
2. Benítez, M., Medina, J., Ślęzak, D.: Reducing information systems considering similarity relations. In: Kacprzyk, J., Koczy, L., Medina, J. (eds.) 7th European Symposium on Computational Intelligence and Mathematics (ESCIM 2015), pp. 257–263 (2015)
3. Benítez-Caballero, M.J., Medina, J., Ramírez-Poussa, E.: Attribute Reduction in Rough Set Theory and Formal Concept Analysis, pp. 513–525. Springer International Publishing, Cham (2017)
4. Cornejo, M.E., Medina, J., Ramírez-Poussa, E.: Irreducible elements in multi-adjoint concept lattices. In: International Conference on Fuzzy Logic and Technology, EUSFLAT, vol. 2013, pp. 125–131 (2013)
5. Cornejo, M.E., Medina, J., Ramírez-Poussa, E.: Attribute reduction in multi-adjoint concept lattices. Inf. Sci. **294**, 41–56 (2015)
6. Davey, B., Priestley, H.: Introduction to Lattices and Order, 2nd edn. Cambridge University Press (2002)

7. Dias, S., Vieira, N.: Reducing the size of concept lattices: the JBOS approach. In: 7th International Conference on Concept Lattices and Their Applications (CLA 2010), vol. 672, pp. 80–91 (2010)
8. Gabbay, D.M., Guenthner, F. (eds.): Handbook of Philosophical Logic. Volume I: Elements of Classical Logic, vol. I. Springer Netherlands (1983)
9. Ganter, B., Wille, R.: Formal Concept Analysis: Mathematical Foundation. Springer (1999)
10. Janusz, A., Ślęzak, D., Nguyen, H.S.: Unsupervised similarity learning from textual data. Fundam. Inf. 119(319–336), 01 (2012)
11. Medina, J.: Relating attribute reduction in formal, object-oriented and property-oriented concept lattices. Comput. Math. Appl. 64(6), 1992–2002 (2012)
12. Pawlak, Z.: Rough sets. Int. J. Comput. Inf. Sci. 11, 341–356 (1982)
13. Ślęzak, D., Janusz, A.: Ensembles of bireducts: towards robust classification and simple representation. In: Kim, T.-H., Adeli, H., Ślęzak, D., Sandnes, F., Song, X., Chung, K.-I., Arnett, K. (eds.) Future Generation Information Technology. Lecture Notes in Computer Science, vol. 7105, pp. 64–77. Springer, Berlin (2011)
14. Stawicki, S., Ślęzak, D.: Recent advances in decision bireducts: complexity, heuristics and streams. In: Lecture Notes in Computer Science, vol. 8171, pp. 200–212 (2013)
15. Wei, L., Qi, J.-J.: Relation between concept lattice reduction and rough set reduction. Knowl. Based Syst. 23(8), 934–938 (2010)

Similarity Measure Between Linguistic Terms by Using Restricted Equivalence Functions and Its Application to Expert Systems

Clemente Rubio-Manzano

Abstract The aim of this paper is to propose a new similarity measure between linguistic terms by using restricted equivalence functions. We formally define it and prove that it fulfills similarity conditions. We explain how this measure can be employed for improving approximate reasoning in expert system inference engines and for maximizing similarity measures between linguistic terms. Finally, an experimental comparison between similarity measures is performed.

1 Introduction

Similarity between elements from a set is an important concept in several disciplines. In machine learning similarity is employed in order to find out regions within the data space with similar features, in computer vision similarity plays an important role in classification, clustering, image segmentation, object tracking and recognition.

Due to the great variety of disciplines there is not exists a universal similarity measure, it depends on the underlying data structure. For example, in social data analytics, similarity is defined on a collection of graphs, in bio-informatics is established on strings, in natural language processing between words and grammars, in fuzzy logic is established on different kind of fuzzy sets, in logic programming, similarity is defined between terms of a first-order language.

In this paper a new similarity measure between linguistic terms based on restricted equivalence functions is proposed. This measure is formally defined and it is proved that it fulfills similarity conditions. An experimental exploration is performed by implementing it in a real system called Bousi Prolog.[1] We show that it is a well-suited measure for maximizing similarity and for improving approximate reasoning in expert system inference engine.

---

[1] Web Site of Bousi Prolog: http://www.face.ubiobio.cl/~clrubio/bousiTools/.

C. Rubio-Manzano (✉)
Department of Information Systems, University of the Bío-Bío, Concepción, Chile
e-mail: clrubio@ubiobio.cl

© Springer Nature Switzerland AG 2019
M. E. Cornejo et al. (eds.), *Trends in Mathematics and Computational Intelligence*, Studies in Computational Intelligence 796,
https://doi.org/10.1007/978-3-030-00485-9_11

The outline of the paper is as follows. In Sect. 2 general concepts regarding linguistic terms and restricted equivalence functions are introduced. In Sect. 3 a similarity measure between linguistic terms using Restricted Equivalent Functions is proposed and formally defined. An experimental evaluation is performed in Sect. 4 and, finally, we give our conclusions and future research lines in Sect. 5.

2 Preliminaries Concepts

2.1 Linguistic Variable and Proximity Equations

A *linguistic variable* is a quintuple $(X, T(X), U, G, M)$ where: X is the variable name, $T(X)$ is the set of linguistic terms of X (i.e., the set of names of linguistic values of X), U is the domain or universe of discourse, G is a grammar that allows to generate $T(X)$ and M is a semantic rule which assigns to each linguistic term x in $T(X)$ its meaning (i.e., a fuzzy subset of U—characterized by its membership function μ_x—).

A linguistic variable can be represented by using a set proximity equations (PEs). A PE has the form $E_i \equiv x \sim y = \alpha_1$, with $x, y \in T(X)$ and $\alpha \in [0, 1]$. Therefore, PEs are defined on the set of linguistic terms, $T(X)$. To this end, we proceed as follows:

Suppose that $T(X) = \{x_i \mid i \in I\}$, where I is a set of.

indexes For each x_i and x_j, with $i, j \in I$, we generate a

proximity equation on $T(X) : \mathcal{R}(x_i, x_j) = \alpha$. The degree α

is calculated by using a similarity measure between $M(x_i)$

and $M(x_j)$

Fuzzy matching functions play an important role in the most of the expert system inference engines. For example, Fuzzy Clips [7] and Bousi Prolog [8] employ fuzzy matching functions in order to implement similarity measures between linguistic terms [5].

Example 1 Suppose a linguistic variable *Age* with $T(Age) = \{young, middle - aged, old\}$, $U = [0, 100]$ and the trapezoidal membership functions for young, middle-aged and old, respectively: $(0, 0, 30, 50), (20, 40, 60, 80), (50, 80, 100, 100)$. Then, a similarity measure can be established between linguistic terms which can be represented by using proximity equations. For example, the following proximity equations are computed by using the fuzzy matching function implemented by the Fuzzy Clips system.

```
young ~ middle=similarity(young,middle)=0.4
young ~ old=similarity(young,old)=0.0
middle ~ old=similarity(middle,old)=0.3
```

2.2 Restricted Equivalent Functions

Restricted equivalence functions (REFs) [1, 2] have been proposed in image processing in order to obtain sequences of optimal thresholds in images with several objects [3].

Definition 1 A REF, f, is a mapping $[0, 1]^2 \longrightarrow [0, 1]$ which satisfies the following conditions:

1. $f(x, y) = f(y, x)$ for all $x, y \in [0, 1]$
2. $f(x, y) = 1$ if and only if $x = y$
3. $f(x, y) = 0$ if and only if $x = 1$ and $y = 0$ or $x = 0$ and $y = 1$
4. $f(x, y) = f(c(x), c(y))$ for all $x, y \in [0, 1]$, c being a strong negation.
5. For all x,y,z $\in [0, 1]$, if $x \le y \le z$, then $f(x, y) \ge f(x, z)$ and $f(y, z) \ge f(x, z)$

For example, $g(x, y) = 1 - |x - y)$ satisfies conditions (1)–(5) with $c(x) = 1 - x$ for all $x \in [0, 1]$. A similarity measure based on REFs between linguistic terms is here proposed in order to enhance the inference engine of Bousi Prolog.

3 Similarity Between Linguistic Terms Using Restricted Equivalence Functions

Firstly, a new similarity measure is proposed. This measure employs a restricted equivalence function in order to compute an approximation degree from the membership values which are calculated for each element of a linguistic term.

Definition 2 Given two linguistic terms A, B defined on the domain $\{u_1, \ldots, u_n\}$ whose membership values are $\mu_A = \{\mu_A(u_1) \ldots, \mu_A(u_n)\}$ and $\mu_B = \{\mu_B(u_1), \ldots, \mu_B(u_n)\}$ respectively. A similarity measure between A and B is defined as:

$$S_{REF}(A, B) = \sum_{i=0}^{n}(REF(\mu_A(u_i), \mu_B(u_i)))/n$$

Secondly, we recall similarity concept proposed in [4]. In this definition, similarity conditions are enumerated.

Definition 3 A mapping $S : A \times B \to [0, 1]$, S is a similarity measure if satisfies the following properties.

1. $0 \leq S(A,B) \leq 1$
2. If $A = B$ then $S(A,B) = 1$
3. $S(A,B) = S(B,A)$
4. If $A \subseteq B \subseteq C$ then $S(A,C) \leq S(A,B)$ and $S(A,C) \leq S(B,C)$

Finally, we prove that S_{REF} is a similarity measure which can be established between linguistic terms. It is formalized with the following proposition.

Proposition 1 *A function S_{REF} is a similarity measure between linguistic terms.*

Proof We are going to prove that S_{REF} satisfies the similarity conditions:

 (i) by definition of REFs;
 (ii) if $A = B$ implies that $\forall\ u_i \in A, B\ \mu_A(u_i) = \mu_B(u_i)$ with $1, \ldots, N$, hence $S_{REF}(\mu_A(u_i), \mu_B(u_i)) = 1$.
(iii) direct by definition;
(iv) A partial order is established, hence $\mu_A(u_i) \leq \mu_B(u_i) \leq \mu_C(u_i)$, what implies that:

1. $S_{REF}(\mu_A(u_i), \mu_B(u_i)) \geq S_{REF}(\mu_A(u_i), \mu_C(i))$ and;
2. $S_{REF}(\mu_B(u_i), \mu_C(u_i)) \geq S_{REF}(\mu_A(u_i), \mu_C(u_i))$ by definition and by the property (5)

Hence, we have:

1. $S_{REF}(\mu_A(u_i), \mu_C(u_i) \leq S_{REF}(\mu_A(u_i), \mu_B(u_i))$ and;
2. $S_{REF}(\mu_B(u_i), \mu_C(u_i)) \leq S_{REF}(\mu_A(u_i), \mu_C(u_i))$

$\square$

Example 2 Suppose three linguistic terms defined by using the membership values:

$A = \{1.0, 0.5, 0.25, 0.0\}$
$B = \{1.0.0.5, 0.25, 0.0\}$
$C = \{1.0, 1.0, 1.0.1.0\}$

Similarity can be computed between linguistic terms A, B and C by using S_{REF}:

$S_{REF}(A, B) = (1, 1, 0.75, 0.5) = 0.8$
$S_{REF}(A, C) = (1, 0.5, 0.25, 0) = 0.4$
$S_{REF}(B, C) = (1, 0.5, 0.5, 0.5) = 0.6$.

The similarity S_{REF} fulfills the similarity conditions:

1. $0 \leq S_{REF}(A, C) \leq S_{REF}(B, C) \leq S_{REF}(A, B) \leq 1.0$
2. Let D=$\{1.0, 0.5, 0.25, 0.0\}$ be a linguistic term, then $S_{REF}(A, D) = 1.0$.
3. $S_{REF}(A, D) = S_{REF}(D, A)$, analogously for the rest of linguistic terms.
4. $A \subseteq B \subseteq C$ what implies that $S_{REF}(A, C) \leq S_{REF}(A, B)$ and $S_{REF}(A, C) \leq S_{REF}(B, C)$

4 Implementation and Evaluation

We have implemented and incorporated this new measure of similarity into the Bousi Prolog system. The linguistic variables of the Example 1 can be programmed in Bousi Prolog as follows:

```
%% linguistic variable
:-domain(age(0,100,years)).
:-fuzzy_set(age,[young(0,0,30,50),
                 middle(20,40,60,80),
                 old(50,80,100,100)]).
%% facts
person ( john , young ).
person ( mary , age#35 ).
person ( paul , middle ).
person ( warren , old ).
```

Then, linguistic variables are compiled in a set of proximity equations together with the rest of the program. The program could be a set of fact representing age's people. We can now compare the original similarity measure (explained in [8]) with the measure S_{REF}. A set of proximity equations generated by using the original similarity measure and a set of proximity equations generated by using S_{REF} is shown in the Fig. 1.

Note that, S_{REF} allows to compute approximation degrees upper than the original one and to found new relationship. For example, it is capable of founding relationship between the linguistic term "young" and "old". Also, the approximation degree between "middle" and "old" is upper than the obtained by the original measure.

Continuing with example, related to the linguistic variable Age and taking the clauses of Fig. 1. If we ask about what people are young, "?-person(X,young).", the BPL system answers:

```
    X=john with 1.0 ;
    X=mary with 0.8 ;
    X=paul with 0.4 ;
    X=warren with 0.2
```

With the incorporation of S_{REF} in the core of the system, it is capable of obtaining more answers.

Fig. 1 Proximity equations versus proximity equations based on REF

Dubois-Prade-Testamale	Restricted equivalence functions
$young \sim age35 = 0.8.$	$young \sim age35 = 0.8.$
$middle \sim age35 = 0.8.$	$middle \sim age35 = 0.8.$
$old \sim age35 = 0.0.$	$old \sim age35 = 0.0.$
$young \sim middle = 0.4.$	$young \sim middle = 0.4.$
$young \sim old = 0.0.$	$young \sim old = 0.2.$
$middle \sim old = 0.3.$	$middle \sim old = 0.4.$

5   Conclusions and Future Work

A new similarity measure between linguistic terms has been proposed. We have formally defined it and proved that it fulfills similarity conditions. A experimental comparison by incorporating it in the core of the Bousi Prolog system has been analyzed. We have shown that restricted equivalence functions can be used as a measure of similarity between linguistic terms which allows us to enhance the Bousi Prolog inference engine.

Acknowledgements This work has been done in collaboration with the research group SOMOS (SOftware-MOdelling-Science) funded by the Research Agency and the Graduate School of Management of the Bío-Bío University.

References

1. Bustice, H., Barrenechea, E., Pagola, M.: Restricted equivalence functions. Fuzzy Sets Syst. **157**, 2333–2346 (2006)
2. Bustice, H., Barrenechea, E., Pagola, M.: Image thresholding using restricted equivalence functions and maximizing the mesaures of similarity. Fuzzy Sets Syst. **158**, 496–516 (2007)
3. Bustice, H., Barrenechea, E., Pagola, M.: Relation between restricted dissimilarity functions, restricted equivalence functions and normal E_N-functions: image thresholding invariant. Pattern Recognit. Lett. **29**, 525–536 (2008)
4. Dengfeng, L., Chuntian, C.: New similarity measures of intuitionistic fuzzy sets and application to pattern recognitions. Pattern Recognit. Lett. **23**(1), 221–225 (2002)
5. Dubois, D., Prade, H., Testemale, C.: Weighted fuzzy pattern matching. Fuzzy Sets Syst. **28**, 313–331 (1988)
6. Julián-Iranzo, P., Rubio-Manzano, C.: An efficient fuzzy unification method and its implementation into the Bousi Prolog system. In: 2010 IEEE International Conference on Fuzzy Systems (FUZZ), pp. 1–8. IEEE
7. Orchard, R.A.: FuzzyClips Version 6.04A. User's Guide Integrated Reasoning. Institute for Information Technology. Canada (1998)
8. Rubio-Manzano, C., Julin-Iranzo, P.: A Fuzzy linguistic prolog and its applications. J. Intell. Fuzzy Syst. Appl. Eng. Technol. **26**(3), 1503–1516 (2014)

A New Convex Hull, Sliding Window Based Online Adaptation Method for Fixed-Structure Radial Basis Function Neural Networks

H. Khosravani, A. Ruano and P. M. Ferreira

Abstract In any online adaptation scheme, two important phenomena should be taken into consideration; parameter shadowing and parameter interference. To alleviate these problems, in this paper a sliding window based online adaptation method for fixed-structure Radial Basis Function Neural Networks (RBFNNs) is proposed. The method is capable of updating the underlying model using the new arriving samples reflecting, to a good extent, new regions in the input-output space and also can deal with the two phenomena mentioned above. The online adaptation process requires a small update rate, while maintaining a good level of accuracy of the updated model.

Keywords Convex hull · Multi objective genetic algorithm
Online adaptation process · Radial basis function neural networks
Time series models

1 Introduction

When a system to be modelled is time-varying (i.e., its dynamics and operating regions change over time), either the model must be retrained or must be adapted online. In order to do this, sequential learning methods, also called online learning

H. Khosravani · A. Ruano (✉)
Faculty of Science and Technology, University of Algarve, Faro, Portugal
e-mail: aruano@ualg.pt

H. Khosravani
e-mail: hkhosravani@csi.fct.ualg.pt

H. Khosravani · A. Ruano
IDMEC, Instituto Superior Técnico, Universidade de Lisboa, Lisbon, Portugal

P. M. Ferreira
LaSIGE, Faculdade de Ciências, Universidade de Lisboa, Lisbon, Portugal
e-mail: pmf@ciencias.ulisboa.pt

© Springer Nature Switzerland AG 2019
M. E. Cornejo et al. (eds.), *Trends in Mathematics and Computational Intelligence*, Studies in Computational Intelligence 796,
https://doi.org/10.1007/978-3-030-00485-9_12

methods, are applied. Regarding the model structure, online learning methods are categorized into two main classes. In the first class, the structure of the model, translated into the number of hidden neurons/layers and the input features employed, is constant over the adaptation process and only the parameters are adjusted [1, 2]. In the second class, focusing only on the model topology, hidden neurons are inserted or removed from the model structure using specific growing and pruning algorithms, respectively [3, 4]. In this paper, a new online adaptation method based on the convex hull concept and a sliding-window technique to update fixed-structure RBFNN models is proposed. This method is an extension of the ones proposed in [2]. The use of the two concepts, convex hull and sliding window, enables the proposed method not only to maintain the previous mappings (i.e., avoiding the parameter interference phenomenon) but also significantly prevents from unnecessary updates (i.e., avoiding the parameter shadowing phenomenon).

This paper is organized as follows. The new method is introduced in Sect. 2. Simulation results obtained from a case study are presented in Sect. 3. A comparison of the performance of the proposed method with other techniques is given in Sect. 4. Finally, conclusions and future work are presented in Sect. 5.

2 Proposed Online Adaptation Method

The idea behind the proposed method is updating the underlying model only if the new arriving sample at any time instant k changes the range of the input-output space. In this case, the current convex hull is updated with the new sample, and included in the training sliding window. The method employs not only a training window, but also an additional sliding window. The size of both sliding windows is constant throughout the online adaptation process and they are managed by a convex hull based policy which benefits from the policy proposed in [2], hereinafter called *F-R policy*. Once the sliding windows are updated, by checking the two update criteria introduced in [2], the model's parameters are adjusted using a modified Levenberg-Marquardt (LM) method. Mainly, the proposed method can be summarized into three steps, discussed below.

2.1 Evaluation of the Arriving Sample

At each sampling instant, a new arriving sample is evaluated to see whether it leads to a new range of input-output space, or not, by comparing it with the current convex hull. The new sample is considered as an informative sample when it is located outside the current convex hull, meaning that a new range of input-output space must be determined, including the new point. The dimension of both the new arriving sample and the convex hull vertices is equal to the number of inputs of the involving RBFNN model plus one, since the model has only one output. To determine whether the new sample is located outside the current convex hull or not, a convex hull algorithm is applied on a set containing the vertices of the current convex hull and the new sample. If the new sample is marked as a new vertex of the convex hull, it is definitely located outside the current convex hull; otherwise, it is considered as an inner point. To locate the new point with respect to the current convex hull, a heuristic along with the *ApproxHull* method, proposed in [5], is applied. The heuristic uses a user-defined distance threshold β to determine whether the new point is likely located outside the convex hull or not.

2.2 Sliding Windows Update

Typically, in sliding windows online algorithms, a FIFO management scheme is used in the management of the window. In the proposed online adaptation method, two management schemes can be applied to the training and the additional sliding windows: one is the F_R policy and the other is a new convex hull based policy proposed in this work. The idea behind the F_R policy is updating the sliding window T with the new sample p if p brings new information to T, while keeping a desirable level of diversity in the window.

When p is presented to the model, two steps should be performed to update T. The first one is whether p should be inserted into T. If so, the second one is which sample of T should be replaced with p, since the size of T is assumed to be constant throughout the online adaptation process. For the first and second point, two criteria called *Include* and *Exclude* are used, respectively. The *Include* criterion checks whether p has enough dissimilarity to all points of T. To do this, the Euclidean distances between p and all points are computed and, if all distances are greater than a user-defined threshold η, point p is inserted into T. The main idea behind the *Exclude* criterion is to randomly remove one of two points in T which have the largest similarity (i.e., the minimum Euclidean distance) among all pairs of points. The proposed convex hull based policy is introduced below.

A Convex Hull Based Policy. If p is considered as an outer point with respect to the current convex hull, the current convex hull is updated considering p as a new vertex. In this step, if some vertices, denoted by *in*, of the current convex hull are marked as inner points by *ApproxHull*, they are replaced with points extracted from the additional sliding window. These points, denoted as *out*, are selected as the ones with the largest dissimilarity to the vertices in the current convex hull. The *in* and *out* points are then swapped between sets. p is inserted into the training window, and the sample to be extracted is determined using the *F-R policy*. If p is not inserted into the training window, it is considered to be inserted into the additional window employing by the *F-R policy*.

From the implementation point of view, the sliding windows can be considered as matrices of size $n_s \times d$ where n_s and d denote the number and the dimension of points, respectively (in the example in this paper, they are 1548×10 and 500×10, for the training and additional sliding windows).

2.3 Parameters Update

The idea behind the procedure of parameters update in this work is the same as that used in [2]. We assume that the change of dynamics of most processes is gradual over time. Hence, the underlying model does not need necessarily to be updated whenever a new sample arrives and is inserted into the training sliding window. Additionally, frequently updating parameters over a period of time not only is translated in an extra computational cost, but also may cause over-training. In the LM method, the training process ends when two termination criteria are met simultaneously:

$$\Phi_k - \Phi_{ku} < \theta_k \qquad (1)$$

$$\|\mathbf{g}_k\| \le \sqrt[3]{\tau_f}(1 + \Phi_k) \qquad (2)$$

In the previous equations, $\theta_k = \tau_f(1 + \Phi_k)$ and Φ_{ku} and Φ_k denote the values of the MSE obtained with the current parameters update and with the previous parameters, respectively. $\mathbf{g}_k$ is the gradient vector of the MSE and τ_f is a resolution parameter denoting a desired correct number of digits in the solution. If one of them is not met, the model parameters are updated by the LM method starting with the parameters obtained in the last update until the two criteria are met. As an alternative terminating criterion, the early-stopping method can be used, using the additional sliding window as a test set.

3 Simulation Results

To evaluate the performance of the proposed online adaptation method, a case study was considered. In this study, a time series Nonlinear Auto Regressive (NAR) model was chosen to compute the one-step-ahead value of the Outside Air Temperature (OAT). The corresponding models were designed offline using one execution of a Multi-Objective Genetic Algorithm (MOGA), discussed in [6]. Regarding the MOGA's parameters, both the maximum number of generations and the population size were set to 100. The early-stopping method was applied with a maximum of 100 iterations. After one complete run of MOGA, one model was selected from the non-dominated set for the case study.

3.1 Case Study: OAT Model for the University of Almeria

The data provided by the University of Almeria has been collected over the years 2010–2012, including climate variables such as outside air temperature, outside air humidity, outside solar radiation, etc. In the design process, the data in the range 02-Sep-2010 to 11-Sep-2010 (i.e., 10 days) with a sample rate of 5 min was used

Table 1 Statistical results of experiments

n_T	n_A	n_R	n_U	n_I	n_{CH}
3306	55509	368	50	5.47	163

to create the training, testing and validation sets with 1548, 516 and 516 points, respectively. *ApproxHull* was applied on the whole data which resulted in 880 convex hull points that were included in the training set. In this process, the range of features considered by MOGA comprised the first 48 lags (i.e., corresponding to the first 4 previous hours), together with 25 lags centered on the sample corresponding to one day before (1 h before and 1 h after). Therefore, 73 features were considered by MOGA, and a RBFNN model with 9 inputs and 14 hidden neurons was selected from the non-dominated set. To simulate the online adaptation process, 12 periods over the years 2010 and 2011 were considered. The samples of each period were normalized in the range $[-1, 1]$. In this case study, two groups of experiments were considered. Each group included three experiments. For all experiments, η (the user-defined threshold used in *Include* criterion of *F-R policy*, please refer to Sect. 2.2) was set to 0.005. For the first and second group, τ_f was set to 0.001 and 0.0001, respectively. For the second group, the early-stopping method was considered. For both groups, β (the user-defined distance threshold used in the heuristic proposed to evaluate the arriving sample, please refer to Sect. 2.1) was taken from $\{0.0, 0.1, 0.5\}$.

For all experiments, the online adaptation process starts with the parameters' values obtained in the offline MOGA design. The model is subsequently updated over the periods in time order. In this procedure, at the beginning of each period, the online adaptation process continues with the last update of the model over the previous period. For all experiments, the sizes of the training and the additional sliding window size were set to 1548 and 500, respectively.

The statistical and evaluation results obtained from the two groups of experiments are given in Tables 1 and 2, respectively. In Table 1, n_T and n_A denote the average number of samples which have been inserted into the training and additional sliding window over all periods for all experiments, respectively. n_R refers to the average number of samples which have been rejected from inserting into both the training and additional sliding windows over all periods for all experiments. n_U and n_I denote the average number of parameter updates and average number of iterations of training process per each update over all periods for all experiments, respectively. n_{CH} denotes the average number of convex hull vertices at the end of the online adaptation process.

As it can be seen in Table 1, n_U is much smaller than n_T. This result reveals the fact that the proposed method can prevent unnecessary parameter updates whenever the training sliding window is updated due to the insertion of the new arriving sample. For all experiments, the initial and updated models have been evaluated over each period. In Table 2, ρ_1^i and ρ_1^u denote the average of the scaled 1-step-ahead RMSE associated with the initial (offline) and the updated model at the end of the period over all experiments, respectively. The bold values in Table 2 refer to the best results.

Table 2 Evaluation results of experiments

	Oct	Nov	Dec	Jan	Feb	Mar	Apr	May	Jun	Jul	Aug	Sep
ρ_1^i	0.075	0.387	0.290	0.480	0.409	0.293	0.152	0.046	0.024	0.007	0.017	0.005
ρ_1^u	0.006	0.007	0.007	0.005	0.008	0.006	0.006	0.007	0.006	0.006	0.006	0.005

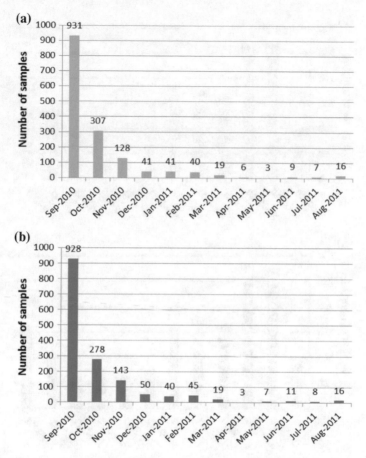

Fig. 1 Comparison of the training sliding windows for the first experiments ($\eta = 0.005$, $\beta = 0.0$) of the first (**a**) and the second (**b**) groups

As it can be seen, while the updated model obtains a good performance throughout the year, this does not happen with the offline model.

To clarify how the sliding window policies perform, the contents of the training sliding window at the end of the period Aug-2011, for both groups of experiments, are shown in Figs. 1 and 2. In these figures, the number of samples (y-axis) entering the sliding window in the corresponding period (x-axis) are shown, i.e., it illustrates how the 1548 samples are decomposed per period. As it can be observed, for each pair of experiments where the same β has been used, the pattern of training sliding window update is similar, resulting in somehow the same sliding window at the end of the last period, Aug-2011.

Moreover, as it can be observed, by increasing β, the update rate of the initial training sliding window containing samples of Sep-2010 is increasing, and gradually, the samples of Sep-2010 are being replaced with the new arriving samples of the other

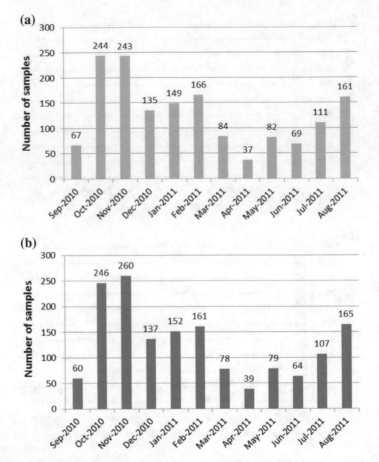

Fig. 2 Comparison of the training sliding windows for the third experiments ($\eta = 0.005$, $\beta = 0.5$) of the first (**a**) and the second (**b**) groups

periods. Since the training sliding window managed by the both policies, involves a diversity of samples over all the periods, it can overcome the common FIFO policy disadvantage, which forgets information of past periods.

4 Comparison with Other Methods

Using the data used in the last section, the results obtained in the best experiment of the proposed method, denoted as CHSWNLM, were compared with those obtained of employing the methods proposed in [2]. As it has been referred, in [2] two methods using a sliding window strategy, called SWNLM and SAWNLM were proposed, and served as the basis of the method introduced in this work. In SWNLM, the sliding

Table 3 Comparison of statistical results obtained by CHSWNLM, SWNLM and SAWNLM

	n_T	n_A	n_R	n_U	n_I
CSWNLM	**414**	58393	377	**24**	4.25
SWNLM	59184	–	0	270	2.77
SAWNLM	58794	–	390	52	3.21

window is managed using FIFO policy, while in the SAWNLM, the sliding window is managed by *F-R policy*. The statistical results of the three methods are shown in Table 3. The total number of new samples presented to the model over the all periods is 59184. The statistics presented in Table 3 are the ones used in Table 1. As it can be seen, the total number of new arriving samples which have been inserted into the training sliding window (n_T), the total number of updates (n_U) and the total number of iterations ($n_U \times n_I$) in CHSWNLM are much smaller than in the other methods. The average of scaled 48-steps-ahead RMSEs over all periods for CHSWNLM, SAWNLM and SWNLM are 0.1219, 0.1339 and 0.575, respectively. It shows that the CHSWNLM has slightly better performance than SAWNLM and SWNLM has the worst performance.

5   Conclusions

In this paper, a sliding window based online adaptation method was proposed to update fixed-structure RBFNN models, previously designed offline. The proposed method is an extension of the one proposed in [2], employing the convex hull concept, incorporating the current sample in the training sliding window if it lies outside the current convex hull. Simulation results showed that the proposed method can significantly improve the performance of offline designed models for time-varying processes. In addition, it presents a performance similar to SAWNLM and has significantly better performance than SWNLM, requiring for that a much smaller number of insertions in the sliding window, number of updates and total number of iterations.

References

1. Ruano, A.E., Crispim, E.M., Conceicao, E.Z.E., Lucio, M.: Prediction of building's temperature using neural networks models. Energy Build. **38**(6), 682–694 (2006)
2. Ferreira, P.M., Ruano, A.E.: Online sliding-window methods for process model adaptation. IEEE Trans. Instrum. Measur. **58**, 3012–3020 (2009)
3. Platt, J.: A resource-allocating network for function interpolation. In: Unsupervised Learning: Foundations of Neural Computation, pp. 341–353 (1999)
4. Lu, Y.W., Sundararajan, N., Saratchandran, P.: A sequential learning scheme for function approximation using minimal radial basis function neural networks. Neural Comput. **9**(2), 461–478 (1997)

5. Khosravani, H.R., Ruano, A.E., Ferreira, P.M.: A convex hull-based data selection method for data driven models. Appl. Soft Comput. **47**, 515–533 (2016)
6. Ferreira, P., Ruano, A.: Evolutionary multiobjective neural network models identification: evolving task-optimised models. In: Ruano, A., Várkonyi-Kóczy, A. (eds.) New Advances in Intelligent Signal Processing. vol. 372, pp. 21–53. Springer, Berlin, Heidelberg (2011)

A Population Based Metaheuristic for the Minimum Latency Problem

Boldizsár Tüű-Szabó, Péter Földesi and László T. Kóczy

Abstract In this paper we present a population based metaheuristic for solving the Minimum Latency Problem, which is the combination of bacterial evolutionary algorithm with local search techniques. The algorithm was tested on TSPLIB benchmark instances, and the results are competitive in terms of accuracy and runtimes with the state-of-the art methods. Except for two instances our algorithm found the best-known solution, and for the biggest tested instance it outperformed the best-known solution. The runtime was on average 30% faster than the most efficient method in the literature.

Keywords Discrete optimization · Minimum latency problem
Delivery man problem · Traveling repairman problem · Metaheuristic

1 Introduction

1.1 The Minimum Latency Problem

The Minimum Latency Problem is a variant of the Traveling Salesman Problem. It also called the Traveling Repairman Problem or the Delivery Man Problem. The task is to find a Hamiltonian circuit that minimalize the sum of arrival times at each

B. Tüű-Szabó (✉) · L. T. Kóczy
Department of Information Technology, Széchenyi István University, Győr, Hungary
e-mail: tszboldi@gmail.com; tuu.szabo.boldizsar@sze.hu

L. T. Kóczy
e-mail: koczy@sze.hu

P. Földesi
Department of Logistics, Széchenyi István University, Győr, Hungary
e-mail: foldesi@sze.hu

L. T. Kóczy
Department of Telecommunications and Media Informatics, Budapest University
of Technology and Economics, Budapest, Hungary

© Springer Nature Switzerland AG 2019
M. E. Cornejo et al. (eds.), *Trends in Mathematics and Computational Intelligence*, Studies in Computational Intelligence 796,
https://doi.org/10.1007/978-3-030-00485-9_13

113

node. The Minimum Latency Problem has also many application areas: logistics, customer-centric routing, scheduling and data retrieval in computer networks.

The Minimum Latency Problem can be defined as a graph search problem with edge weights (1):

$$G_{MLP} = (V_{cities}, E_{conn})$$

$$V_{cities} = v_0 U\{v_1, v_2, \ldots, v_n\}, E_{conn} \subseteq \{(v_i, v_j)|i \neq j\} \tag{1}$$

$$C : V_{cities} \times V_{cities} \rightarrow R, C = (c_{ij})_{(n+1) \times (n+1)}$$

C is called cost matrix, where c_{ij} the cost of going from vertex i to vertex j.

The goal is to find a permutation of vertices $(p_1, p_2, p_3, \ldots, p_n)$ that minimalizes the sum of arrival times (C_{sum}) at each node (2).

arrival time at p_i node: $C_{p_i} = C_{v_o,p_1} + \sum_{j=2}^{i} C_{p_{j-1},p_j}$ $i = 1\ldots n$

$$C_{sum} = \sum_{i=1}^{n} C_{p_i} \tag{2}$$

1.2 Our Previous Work

In recent years we compared various population based algorithms (genetic algorithm [4], bacterial evolutionary algorithm [9], particle swarm algorithm [5] and their memetic versions [2, 3]).

In 2016 we presented a Discrete Bacterial Memetic Evolutionary Algorithm (DBMEA) for The Traveling Salesman Problem [6, 7]. The algorithm showed good properties: it founded optimal and near-optimal solutions for instances up to 1000 vertices. In 2016 we presented an improved Discrete Bacterial Memetic Evolutionary Algorithm with bounded local search which led to significant improvement in runtime [12].

Considering the above we came to the conclusion is worth to examine our algorithm on other TSP variants.

2 Our Method

Our method is called Discrete Bacterial Memetic Evolutionary Algorithm (DBMEA). The DBMEA is a memetic algorithm, a combination of the bacterial evolutionary algorithm and 2-opt, 3-opt local search. The process of the DBMEA algorithm can be seen in Fig. 1.

Fig. 1 The process of
DBMEA algorithm

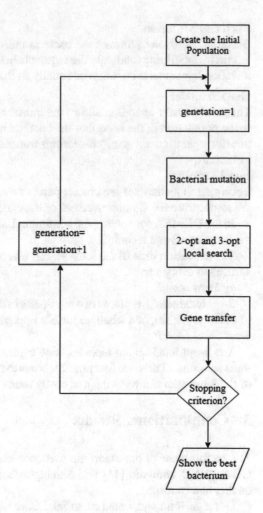

Memetic algorithms combine the global search evolutionary algorithms with local search methods, so eliminate the disadvantages of both methods [8], so in many cases the addition of local search approaches usually can improve significantly the performance of the classical evolutionary algorithms (both the best solution and the convergence speed).

Creating the Initial Population

The Population is a group of individuals which mean solutions for the problem. In the DBMEA algorithm the individuals represent possible tours for the Minimum Latency Problem.

In the DBMEA for Minimum Latency Problem the individuals in the initial Population are generated randomly.

Bacterial Mutation

Bacterial mutation optimizes the bacteria individually. It tries to improve the bacterium by modifying randomly the segments in the clones. As a result of the bacterial mutation the bacteria are more or equally fit than the original bacteria.

Gene Transfer

The gene transfer operation allows the transfer of information between the bacteria in the population in the hope that the bacteria become better and better. In the gene transfer operation a "good" bacterium transfers a part of its solution to a "bad" bacterium.

Local Search

Local search techniques are crucial parts of every memetic algorithm because they are responsible for the improvement of the candidate solutions in the population.

In the DBMEA for solving the Minimum Latency Problem the local searches are the 2-opt and 3-opt techniques.

The execution time of the local search was reduced by examining only the close vertices to every vertex.

2-opt local search

2-opt local search replaces two edge pairs in the original graph to reduce the length of the tour. It is stopped when no further improvement is possible.

3-opt local search

The 3-opt local search replaces edge triples. The deleting of three edges results three sub-tours. There are four possible ways to reconnect these sub-tours. The output of the 3-opt step is always the less costly tour.

3 Computational Results

The performance of our algorithm was compared with the state-of-the art methods, GILS-RVND heuristic [11] and Salehipour's method [10] in terms of solution's quality and runtime.

Our algorithm was tested on an Intel Core i7-7500U 2.7 GHz, 8 GB RAM workstation. The GILS-RVND was executed an Intel Core i7 2.93 GHz, with 8.0 GB of RAM memory. Salehipour's method was tested on Pentium 4, 2.4 GHz processor and 512 MB RAM.

In Table 1 the results for TSPLIB instances selected by Abeledo et al. [1] can be seen. Our algorithm found the best-known solution for all instances, and the average runtime was shorter than in the case of GILS-RVND heuristic.

In Tables 2 and 3 it can be seen the comparison of DBMEA algorithm with GILS-RVND heuristic and Salehipour's method on TSPLIB instances selected by Salehipour et al. in terms of solution's quality and runtimes. Expect for two instances (lin318 and pr439) our method found the best-known solution, and for the biggest tested instance (att532) it found even a better solution than the best-known. The average solution (averaging 10 runs) was in the case of all instances the same or shorter than in the case of GILS-RVND heuristic. The DBMEA was much faster than the two other methods.

Table 1 Results for TSPLIB instances selected by Abeledo et al.

Instance	Best known	DBMEA			GILS-RVND		
		Best value	Avg. value	Avg. Sec	Best value	Avg. value	Avg. Sec
dantzig42	12528	12528	12528	0.19	12528	12528	0.16
swiss42	22327	22327	22327	0.13	22327	22327	0.16
att48	209320	209320	209320	0.76	209320	209320	0.32
gr48	102378	102378	102378	0.57	102378	102378	0.33
hr48	247926	247926	247926	0.38	247926	247926	0.3
eil51	10178	10178	10178	1.01	10178	10178	0.49
berlin52	143721	143721	143721	1.13	143721	143721	0.46
brazil58	512361	512361	512361	0.56	512361	512361	0.78
st70	20557	20557	20557	2.26	20557	20557	1.65
eil76	17976	17976	17976	2.01	17976	17976	2.64
pr76	3455242	3455242	3455242	1.64	3455242	3455242	2.31
rat99	57986	57986	57986	10.28	57986	57986	11.27
kroA100	983128	983128	983128	5.24	983128	983128	8.59
kroB100	986008	986008	986008	6.92	986008	986008	9.21
kroC100	961324	961324	961324	4.83	961324	961324	8.17

(continued)

Table 1 (continued)

Instance	Best known	DBMEA			GILS-RVND		
		Best value	Avg. value	Avg. Sec	Best value	Avg. value	Avg. Sec
kroD100	976965	976965	976965	5.73	976965	976965	8.46
kroE100	971266	971266	971266	5.66	971266	971266	8.31
rd100	340047	340047	340047	7.60	340047	340047	8.52
eil101	27513	27513	27517.2	16.64	27513	27513	12.76
lin105	603910	603910	603910	7.59	603910	603910	8.42
pr107	2026626	2026626	2026626	6.41	2026626	2026626	10.89
Average				4.50			4.96

Table 2 Solutions for TSPLIB instances selected by Salehipour et al.

Instance	Best known	DBMEA		GILS-RVND		Salehipour et al.	Gap [%] GILS-RVND		Gap [%] Salehipour et al.
		Best value	Avg. value	Best value	Avg. value	Best value	Best value	Avg. value	Best value
st70	19215	19215	19215	19215	19215	19553	0.00	0.00	−1.73
rat99	54984	54984	54984	54984	54984	56994	0.00	0.00	−3.53
kroD100	949594	949594	949594	949594	949594	976830	0.00	0.00	−2.79
lin105	585823	585823	585823	585823	585823	585823	0.00	0.00	0.00
pr107	1980767	1980767	1980767	1980767	1980767	1983475	0.00	0.00	−0.14
rat195	210191	210191	210285.7	210191	210335.9	213371	0.00	−0.02	−1.49
pr226	7100308	7100308	7100308	7100308	7100308	7226554	0.00	0.00	−1.75
lin318	5560679	5562148	5565243.1	5560679	5569820	5876537	0.03	−0.08	−5.35
pr439	17688561	17693137	17707037.33	17688561	17734922	18567170	0.03	−0.16	−4.71
att532	5581240	**5578872**	5584786.2	5581240	5597867	18448435[a]	−0.04	−0.23	−69.76

[a]calculate Euclidean distances instead of ATT pseudo-Euclidean distances

Table 3 Runtimes for TSPLIB instances selected by Salehipour et al.

Instance	DBMEA	GILS_RVND	Salehipour et al.
	Avg. sec	Avg. sec	Avg. sec
st70	1.91	1.51	2.23
rat99	9.87	9.47	9
kroD100	5.29	6.9	11.02
lin105	4.38	6.19	12.23
pr107	4.29	8.13	3.33
rat195	70.20	75.56	311.97
pr226	43.37	59.05	239.56
lin318	268.53	220.59	455.6
pr439	421.75	553.74	5614.74
att532	1078.67	1792.61	5005.32
Average	190.83	273.38	1166.50

4   Conclusions

In this paper an evolutionary metaheuristic was presented for solving the Minimum
Latency Problem. The algorithm is efficient because except for two instances found
the best-known values, for the biggest tested instance it found even a better solution
than the best-known value and the average runtime was smaller than in the case of
the state-of-the-art methods for the problem.

In our further work we plan to test DBMEA algorithm on other TSP variants (time
dependant TSP, multi-TSP etc.).

Acknowledgements This research was supported by the National Research, Development and
Innovation Office (NKFIH) K108405 and by the EFOP-3.6.2-16-2017-00015 "HU-MATHS-IN-
Intensification of the activity of the Hungarian Industrial Innovation Service Network" grant.

MINISTRY
OF HUMAN CAPACITIES

Supported by the ÚNKP-17-3 New National Excellence Program of the Ministry of Human Capacities.

References

1. Abeledo, H., Fukasawa, R., Pessoa, A., Uchoa, E.: The time dependent traveling salesman problem: polyhedra and algorithm. Math. Program. Comput. **5**(1), 27–55 (2013)
2. Botzheim, J., Cabrita, C., Kóczy, L.T., Ruano, A.E.: Fuzzy rule extraction by bacterial memetic algorithms. In: Proceedings of the 11th World Congress of International Fuzzy Systems Association, IFSA 2005, Beijing, China, pp. 1563–1568 (2005)
3. Farkas, M., Földesi, P., Botzheim, J., Kóczy, T.L.: Approximation of a modified traveling salesman problem using bacterial memetic algorithms. In: Towards Intelligent Engineering and Information Technology. SCI vol. 243, pp. 607–625. Springer, Berlin, Heidelberg (2009)
4. Holland, J.H.: Adaption in Natural and Artificial Systems. The MIT Press, Cambridge (1992)
5. Kennedy, J., Eberhart, R.: Particle swarm optimization. In: Proceedings of the IEEE International Conference on Neural Networks (ICNN 1995), Perth, WA, Australia, vol. 4, pp. 1942–1948 (1995)
6. Kóczy, L.T., Földesi, P., Tüű-Szabó, B.: An effective discrete bacterial memetic evolutionary algorithm for the traveling salesman problem. Int. J. Intell. Syst. (2017)
7. Kóczy, L.T., Földesi, P., Tüű-Szabó, B.: A discrete bacterial memetic evolutionary algorithm for the traveling salesman problem. In: IEEE World Congress on Computational Intelligence (WCCI 2016), Vancouver, Canada, pp. 3261–3267 (2016)
8. Moscato, P.: On Evolution, Search, Optimization, Genetic Algorithms and Martial Arts - Towards Memetic Algorithms. Technical Report Caltech Concurrent Computation Program, Report. 826, California Institute of Technology, Pasadena, USA (1989)
9. Nawa, N.E., Furuhashi, T.: Fuzzy system parameters discovery by bacterial evolutionary algorithm. IEEE Tr. Fuzzy Syst. **7**, 608–616 (1999)
10. Salehipour, A., Sörensen, K., Goos, P., Bräysy, O.: Efficient GRASP+VND and GRASP+VNS metaheuristics for the traveling repairman problem. 4OR: A Q. J. Oper. Res. **9**(2), 189–209 (2011)
11. Silva, M.M., Subramanian, A., Vidal, T., Ochi, L.S.: A simple and effective metaheuristic for the minimum Latency problem. Eur. J. Oper. Res. **221**(3), 513–520 (2012)
12. Tüű-Szabó, B., Földesi, P., Kóczy, T.L.: Improved discrete bacterial memetic evolutionary algorithm for the traveling salesman problem. In: Proceedings of the Computational Intelligence in Information Systems Conference (CIIS 2016), Bandar Seri Begawan, Brunei, pp. 27–38 (2017)

Multi-objective Fuzzy Geometric Programming Problem Using Fuzzy Geometry

Debjani Chakraborty, Abhijit Chatterjee and Aishwaryaprajna

Abstract This paper proposes a methodology to obtain fuzzy Pareto optimal frontier of multi-objective fuzzy geometric programming problem. The method that we derive here does not depend on the degree-of-difficulty of the problem under consideration. A study on the fuzzy convexity of the posynomials involved is also given here. Fuzzy geometric and algebraic approaches have been considered for the said optimization problem and is supported with a numerical example.

Keywords Fuzzy convexity · Fuzzy monomial · Fuzzy posynomial
Geometric programming problem · Multi-criteria decision making
Fuzzy geometry

1 Introduction

Geometric programming problem (GPP), a type of mathematical optimization, characterized by objective and constraint functions of a special form was introduced by Duffin et al. [5]. Several engineering applications [3, 9, 10, 12–14] have investigated the effectiveness and importance of geometric programming. Effective algorithms have been developed for solving geometric programming problems [11, 15, 16, 18]. To tackle the uncertainty factors, probabilistic approaches have received attention and stochastic geometric programming has been evolved. Recently in accordance with the advancements of the theory of fuzzy sets and applications, fuzzy geometric programming models have been introduced by Cao [2] in 1987. Geometric programming is a tool to solve a special class of nonlinear programming problems where objective function and constraints are posynomials. Shivanian and Khorram [17] have proposed monomial geometric programming subject to fuzzy relation inequalities. In 2016, Zhou et al. [19] also studied a class of posynomial GPP that considers the minimization of a posynomial subject to fuzzy relational equations with max-min composition.

D. Chakraborty (✉) · A. Chatterjee · Aishwaryaprajna
Indian Institute of Technology, Kharagpur 721302, India
e-mail: drdebjanic@yahoo.co.in; debjani@maths.iitkgp.ac.in

© Springer Nature Switzerland AG 2019
M. E. Cornejo et al. (eds.), *Trends in Mathematics and Computational Intelligence*, Studies in Computational Intelligence 796,
https://doi.org/10.1007/978-3-030-00485-9_14

123

In this paper a methodology has been suggested to solve fuzzy multi objective GPP. The advantage of the proposed methodologies is that, it can capture the solutions even for discrete objective function and/or constraint sets. The degree-of-difficulty (termed as DD) of a GPP is defined as *total number of terms—(total number of variables + 1)*. This characteristic measure plays an important role when we solve a GPP. The larger the DD, the harder the problem to solve. If the DD of a problem is zero, it is possible that the problem can be solved readily. So one always tries to reduce the DD as much as one can by applying his/her engineering knowledge and common sense (Beightler and Phillips [1]). The proposed fuzzy methodology can work efficiently even for negative degree of difficulty.

2 Fuzzy Posynomial

Let us consider the fuzzy posynomial $\tilde{f}(X)$ as

$$\tilde{f}(X) = \sum_{j=1}^{m} \tilde{c}_j \prod_{i=1}^{m} x_i^{a_{ij}} = \sum_{j=1}^{m} U_j \ (say) \tag{1}$$

where $X = (x_1, x_2, ..., x_n)$ is a decision variable vector of the optimization problem and $\tilde{c}_j = (\underline{c}_j, c_j, \bar{c}_j)_{LR}$ is an LR type fuzzy number for all j. Here $\tilde{c}_j$s are all positive fuzzy numbers and the exponents a_{ij}s are real constants (positive, negative or zero), $\forall i, j$. And the decision variables x_is, $\forall i$ are non-negative.

In general, the above mentioned posynomial $\tilde{f}(x)$ is not convex. In optimization set up, the notion of convexity is important to ensure global optimality. Thus the study of convexity is necessary for all posynomial functions. A posynomial is a combination of a monomials. If each term of posynomials, i.e. the monomials are convex, then the posynomial is convex. For fuzzy GPP of minimization type if the posynomial objective function and the constraint set having posynomial functions are convex, then the obtained optimal solution is global optimal. Further generalization may be possible for negative fuzzy $\tilde{c}_j$s and for different cases of a_{ij}s, which in general may be named as fuzzy signomial.

3 Solving Fuzzy Multi-objective GPP

The methodology described in this section is used to solve fuzzy multi-objective GPP Let us consider a general fuzzy GPP as

$$Minimize \, \tilde{f}_k(x) = \sum_{j=1}^{m_k} \tilde{c}_{jk} \prod_{i=1}^{m} x_i^{a_{ijk}} = \sum_{j=1}^{m_k} \tilde{U}_{jk}(say), \ \ k = 1, 2, ...p \tag{2}$$

$$subject \ to \, \tilde{g}_l(x) \gtrsim 0, \ l = 1, 2, 3, ..., q \ x \in \tilde{X}$$

where $\widetilde{c}_{jk} \equiv (c_{jk}^L, c_{jk}, c_{jk}^U)_{LR}$, $\forall k$ are LR type fuzzy numbers, a_{ijk}s are all real numbers and $\widetilde{g}_L(x)$s, $\forall k$ are of posynomial types like $\widetilde{f}_k(x)$s.

Here we are concerned about the pareto-optimal solutions in decision feasible space $\widetilde{X}$ of the multi objective problem. Let $\widetilde{Y}$ be the criteria feasible region for (8). To obtain the non-dominated points of $\widetilde{Y}(1))$ (for $\alpha = 1$), fuzzy ideal cone method (Ghosh and Chakraborty [8]) is needed to be restructured using fuzzy geometry [4, 6, 7] as the feasible space as well as the criterion space is posynomial and so piece-wise continuous. The fuzzy posynomial may be discrete or continuous. For example, $\widetilde{f}(x) = x_1^2 x_2^{-3} + \widetilde{4} x_1^{-2} x_2^5$. It is obvious that the fuzzy function is undefined at $(x_1, x_2) = (0, 0)$, but in all other points the function is continous. The geometry of any fuzzy function can be realized using Ghosh and Chakraborty [6, 7] and Chakraborty and Ghosh [4]. If the fuzzy function is continuous classical optimization technique is applicable to obtain extremum of fuzzy posynomial. But when fuzzy posynomial is piecewise continuous then getting fuzzy derivative is quite impossible in the points of discontinuities. The algorithmic implementation of the following methodology may be suggested for obtaining extremum of a fuzzy posynomial of any kind.

Capturing Complete Fuzzy Non-dominated Set: Let $\widetilde{Y}$ be the criteria feasible region (Ghosh and Chakraborty [8]) for (2). To obtain the non-dominated points of $\widetilde{Y}(1)$ (for $\alpha = 1$), fuzzy ideal cone method is needed to be restructured and solved (corresponding to a particular unit vector $\hat{\beta} \in S_{\geqq}^{k-1} = S^{k-1} \bigcap R_{\geqq}^k$ (where S^{k-1} represents the unit ball in R^k)). In general, a point x^* is said to be pareto optimal point if it is feasible and satisfies the following condition:

$$\left(f(x^*) - R_{\geqq}^{k-1} \right) \bigcap \quad f(X) = \{(f(x^*)\} \tag{3}$$

The geometrical interpretation can be made as—if the objective feasible region and the translated non-positive orthant, whose vertex is being shifted from origin to the point $f(x^*)$, intersect in a single point $f(x^*)$ only, then the feasible point x^* is a pareto optimal solution of the considered multi objective problem. In order to get a non-dominated solution, one may translate the cone of non-positive orthant of the objective space along a particular direction $\hat{\beta} \in R_{\geqq}^k$ till this cone does not touch the objective feasible region. Translation of the cone $-R_{\geqq}^k$ along a particular direction $\hat{\beta} \in R_{\geqq}^k$ means that the vertex of the cone is retained on the line $z\hat{\beta}, z \in R$. Now if the cone is being translated along $\hat{\beta} \in R_{\geqq}^k$ then either the vertex of the cone touches (generates proper pareto optimal point) or the boundary planes may touch the objective feasible region (which generate either weak pareto points or non proper pareto points).

Here the cone $\mathbf{R}^k_{\succeq}$ is used to capture non-dominated points. Geometrically, $z\hat{\beta}$ for $z \geq 0$ represents points on the line which is directed along β and passing through origin. For a particular $z \geq 0$, the constraints of the fuzzy multi objective problem is $\{x \in \widetilde{X}(1) : z\hat{\beta} \succeq \widetilde{f}(x)\}$. To obtain the fuzzy non-dominated set $\widetilde{Y}_N$, which is a subset of the boundary of the criteria feasible region $\widetilde{Y}$, the fuzzy ideal cone methodology is applied to the fuzzy multi-objective GPP described above. A uniform discretization of the set $S^k_{\succeq} = s^{k-1} \bigcup \mathbf{R}^k_{\succeq}$ is considered for the implementation of the fuzzy ideal cone method.

Algorithm 1 Algorithm to solve bi-variate bi-objective fuzzy geometric programming problem

---

Require: Given problem: Given a satisfaction level $\alpha \in [0, 1]$, $\widetilde{X}_\alpha$, i.e. α cut of the feasible space $\widetilde{X}$ is discretized in n number of grid points, where l and k are the indices of the array $\widetilde{X}_\alpha$ such that $l, k = [1, n]$.

    Set m(Number of partition of θ)
    Set n (Number of grid point)
    for $\theta = 0$ to $\frac{\pi}{2}$, step length $\frac{\pi}{2m}$
        Set min (a large value)
        for $l = 1$ to n with step length required
            for $k = 1$ to n with step length required
                if $\widetilde{g}_{\alpha[l,k]} \geq 0$, then
                set $z = \frac{\widetilde{f}_{1\alpha[l,k]}}{cos\theta}$
                    if $z = \frac{\widetilde{f}_{2\alpha[l,k]}}{sin\theta}$, then
                        if $z < min$, then
                            set min=z
                            store value of l and k
                        end if
                    end if
                end if
            end for
        end for
    end for

Algorithmic Implementation to Solve Fuzzy Discrete Multi-objective Geometric Programming Problem: Now we uniformly discretize each θ_i to equal number of subintervals. let we divide θ_1 by m number of point and θ_i by $round(m \prod_{l=1}^{i} sin\phi_l)$ number of points, for $i = 1, 2, ..., (k-1)$. Now for every point of θ_i we obtain min z for the transformed problem of the original GPP. The closed-convex cone $-\mathbf{R}^k_{\succeq}$, which is a non-positive orthant, is shifted along the direction of a particular unit vector $\hat{\beta} \in S^k_{\succeq}$. Then it can be expressed as

$$(\cos\theta_1, \cos\theta_2 \sin\theta_1, \cos\theta_3 \sin\theta_2 \sin\theta_1, ..., \cos\theta_{k-1} \prod_{i=1}^{k-2} \sin\theta_i, \prod_{i=1}^{k-1} \sin\theta_i),$$

where θ_i is restricted between $0°$ to $90°$ and $i - 1, 2, ...(k - 1)$ according to a well known spherical discretization technique [8]. While translating the cone $-\mathbf{R}^k_{\geqq}$ the vertex of the cone is restrained to stay on the line having magnitude z along $\hat{\beta}$, that is $z\hat{\beta}$ where $z \in \mathbf{R}$. The vertex traces the whole criteria space point by point for different α-levels. For a particular α, say α_0 if $z\hat{\beta}$ touches the boundary of the α_0 cut of the criteria space, the point is declared as a member of the non-dominated frontier. Using this concept, the problem (2) can be mathematically restructured in the following manner,

$$Minimize \quad z, \quad subject\ to \quad z(\cos\theta_1, \cos\theta_2 \sin\theta_1...) \geqq [\widetilde{f}(x)]_\alpha, x \in \widetilde{X} \quad (4)$$

where $\widetilde{X}$ is the decision feasible set and $[\widetilde{f}(x)]_\alpha = ([\widetilde{f_1}(x)]_\alpha, [\widetilde{f_2}(x)]_\alpha, ..., [\widetilde{f_k}(x)]_\alpha)$.

Solving the above non-linear programming problem in a discrete or piecewise continuous domain is not possible since a posynomial contains a number of singularities, the nature of the criteria space is piecewise continuous. The following Algorithm 1 is needed to search the whole domain to obtain the non-dominated frontier.

4 Numerical Example

Let us consider the following bi-objective numerical problem where two decision variable are considered. The fuzzy bi-objective GP problem is as follows:

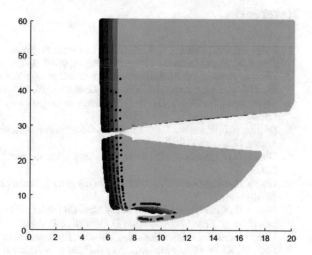

Fig. 1 This is the non-dominated set of the problem

$$Minimize f_1 = x^2 + (0, 2, 4)_{LR}, \quad f_2 = y^2 + (6, 7, 8)_{LR}$$
$$subject\ to (1, 2, 3)_{LR} x^3 y^{-1} \le 10, \quad (2, 3, 4)_{LR} x^{-2} y^{-1} \le 1$$
$$2.7 x^{-\frac{1}{3}} y^2 + x^{\frac{2}{5}} y^{-4} + xy \le 250$$
$$x^2 (y - 5)^{-2} + (x - 1)^{-1.5} \le 30 \tag{5}$$
$$(x - 1.5)^2 (y - 1.5)^2 \ge .5$$
$$x, y \ge 0$$

Now we solve this problem using the proposed algorithm and get the complete fuzzy non-dominated set Fig. 1.

5 Conclusion

Here a methodology has been suggested to obtain solution of multi-objective fuzzy geometric programming problem using fuzzy geometry. The method is robust in nature as it works well even for negative degree of difficulty and efficient for discrete decision space. This method may further be extended for fuzzy signomial programming problems.

Acknowledgements First two authors gratefully acknowledge the financial support provided by Department of Science & Technology, India (SR/S4/MS:858/13).

References

1. Beightler, C.S., Phillips, D.T.: Applied Geometric Programming. Wiley, New York (1976)
2. Cao, B.Y.: Fuzzy Geometric Programming. Applied Optimization, Springer (2002)
3. Choi, J., Bricker, D.: Effectiveness of geometric programming algorithm for optimization of machining economics models. Comput. Oper. Res. **10**, 957–961 (1996)
4. Chakraborty, D., Ghosh, D.: Analytical fuzzy plane geometry II. Fuzzy Sets Syst. **243**, 84–109 (2014)
5. Dufflin, R.J., Peterson, E.L., Zener, C.: Geometric Programming: Theory and Applications. Willey, New York (1967)
6. Ghosh, D., Chakraborty, D.: Analytical fuzzy plane geometry I. Fuzzy Sets Syst. **209**, 66–83 (2012)
7. Ghosh, D., Chakraborty, D.: Analytical fuzzy plane geometry III. Fuzzy Sets Syst. **283**, 83–107 (2016)
8. Ghosh, D., Chakraborty, D.: On Fuzzy Ideal Cone Method to Capture Entire Fuzzy Non dominated Set of Fuzzy Multi-criteria Optimization Problems with Fuzzy Parameters. Facets of Uncertainties and Applications, pp. 249–260. Springer, India (2015)
9. Jung, H., Klein, C.: Optimal inventory policies under decreasing cost functions via geometric programming. Eur. J. Oper. Res. **132**, 628–642 (2001)
10. Kim, D., Lee, W.: Optimal join pricing and lot sizing with fixed and variable capacity. Eur. J. Oper. Res. **109**, 212–227 (1998)

11. Kortanek, K., Xu, X., Ye, Y.: An infeasible interior-point algorithm for solving primal and dual geometric programs. Math. Programm. **76**, 155–181 (1997)
12. Liu, S.T.: Geometric programming with fuzzy parameters in engineering optimization. Internat. J. Approx. Reason. **46**(3), 484–498 (2007)
13. Lui, S.: Posynomial geometric programming with interval exponents and coefficients. Eur. J. Oper. Res. **186**, 17–27 (2008)
14. Qu, S., Zhang, K., Wang, F.: A global optimization using linear relaxation for generalized geometric programming. Eur. J. Oper. Res. **190**, 345–356 (2008)
15. Peterson, E.: The fundamental relations between geometric programming duality, parametric programming duality and ordinary Lagrangian duality. Ann. Oper. Res. **105**, 109–153 (2001)
16. Rajgopal, J., Bricker, D.: Solving posynomial geometric programming problems via generalized linear programming. Computat. Optim. Appl. **21**, 95–109 (2002). World Appl. Sci. J. **7**(1): 94–101 (2009)
17. Shivanian, E., Khorram, E.: Monomial geometric programming with fuzzy relation in equality constraints with max-product composition. Comput. Ind. Eng. **56**, 1386–1392 (2009)
18. Yang, H., Bricker, D.: Investigation of path-following algorithms for signomial geometric programming problems. Eur. J. Oper. Res. **103**, 230–241 (1997)
19. Zhou, X.G., Yang, X.P., Cao, B.Y.: Posynomial geometric programming problem subject to max-min fuzzy relation equations. Inf. Sci. **328**, 15–25 (2016)

Recommendation Solution for a Locate-Based Social Network via Formal Concept Analysis

Jesús Medina, Kristina Pakhomova and Eloísa Ramírez-Poussa

Abstract This paper introduces a procedure to apply Formal Concept Analysis (FCA) to a database obtained from a locate-base social network. In this way, we can know the interest of a target user and make recommendations according to these interests.

1 Introduction

In recent years, data analysis has become a very important and useful research topic in many sectors. One interesting task that we can perform from the information contained in the databases is the creation of a recommendation system. Recommendation systems have been used in many areas such as movies, books, news, social networks, etc. Locate-base social network (LBSN) is one type of Social Network, which allows to share the activities one person is doing via geo-tagged user-generated multimedia content [10]. The LBSN consists of two components: Social Network (SN) and Location. The classical SN allows the users exchange the information, fix a meeting, post the information which is interesting for him/her. The Location component includes the name, the Id, longitude and latitude of each place which is visited by the user.

On the other hand, Formal Concept Analysis (FCA) is a powerful tool for data analysis [1, 2, 4, 5, 8, 9]. The purpose of this mathematical tool is to extract infor-

Partially supported by the State Research Agency (AEI) and the European Regional Development Fund (ERDF) project TIN2016-76653-P.

J. Medina · E. Ramírez-Poussa
Department of Mathematics, University of Cádiz, Cádiz, Spain
e-mail: jesus.medina@uca.es

E. Ramírez-Poussa
e-mail: eloisa.ramirez@uca.es

K. Pakhomova (✉)
Institute of Space and Information Technologies, Siberian Federal University,
Krasnoyarsk, Russia
e-mail: kpahomova@yandex.ru

© Springer Nature Switzerland AG 2019
M. E. Cornejo et al. (eds.), *Trends in Mathematics and Computational Intelligence*, Studies in Computational Intelligence 796,
https://doi.org/10.1007/978-3-030-00485-9_15

131

mation from databases that contain objects and attributes related between them. An interesting generalization of this theory in the fuzzy case was given in [6, 7], where the authors consider the philosophy of the multi-adjoint paradigm in order to provide more flexibility to this theory. In this work, we will consider this fuzzy generalization of FCA to study the interests and habits of one user in order to carry out a recommendation process, considering the information obtained from a LBSN.

2 Preliminaries

In this section, we recall some preliminary definitions related to FCA. The first one is the notion of adjoint triple since they are used to define the concept-forming operators [3].

Definition 1 Let $(P_1, \leq_1)$, $(P_2, \leq_2)$, $(P_3, \leq_3)$ be posets and $\&: P_1 \times P_2 \to P_3$, $\swarrow: P_3 \times P_2 \to P_1$, $\nwarrow: P_3 \times P_1 \to P_2$ be mappings, then $(\&, \swarrow, \nwarrow)$ is an *adjoint triple* with respect to P_1, P_2, P_3 if the following doble equivalence holds:

$$x \leq_1 z \swarrow y \quad \text{iff} \quad x \& y \leq_3 z \quad \text{iff} \quad y \leq_2 z \nwarrow x \tag{1}$$

where $x \in P_1$, $y \in P_2$ and $z \in P_3$.

In the environment of FCA, the posets $(P_1, \leq_1)$ and $(P_2, \leq_2)$ should be complete lattices [7]. Now, the notion of multi-adjoint frame is recalled.

Definition 2 A *multi-adjoint frame* $\mathcal{L}$ is a tuple

$$(L_1, L_2, P, \preceq_1, \preceq_2, \leq, \&_1, \swarrow^1, \nwarrow_1, \ldots, \&_n, \swarrow^n, \nwarrow_n)$$

where $(L_1, \preceq_1)$ and $(L_2, \preceq_2)$ are complete lattices, $(P, \leq)$ is a poset and, for all $i \in \{1, \ldots, n\}$, $(\&_i, \swarrow^i, \nwarrow_i)$ is an adjoint triple with respect to L_1, L_2, P. Multi-adjoint frames are denoted by $(L_1, L_2, P, \&_1, \ldots, \&_n)$.

Once a frame is fixed, we need to consider a formal context.

Definition 3 A *context* is a tuple (A, B, R, σ) such that A and B are non-empty sets (usually interpreted as attributes and objects, respectively), R is a P-fuzzy relation $R: A \times B \to P$ and $\sigma: A \times B \to \{1, \ldots, n\}$ is a mapping which associates any element in $A \times B$ with some particular adjoint triple in the frame.

From a multi-adjoint frame and a context, the concept-forming operators $\uparrow_\sigma: L_2^B \longrightarrow L_1^A$ and $\downarrow^\sigma: L_1^A \longrightarrow L_2^B$ are defined by:

$$g^{\uparrow_\sigma}(a) = \inf\{R(a, b) \swarrow^{\sigma(a,b)} g(b) \mid b \in B\} \tag{2}$$

$$f^{\downarrow^\sigma}(b) = \inf\{R(a, b) \nwarrow_{\sigma(a,b)} f(a) \mid a \in A\} \tag{3}$$

for all $g \in L_2^B$, $f \in L_1^A$ and $a \in A$, $b \in B$.

Taking into account these operators, a *multi-adjoint concept* is a pair $\langle g, f \rangle$ such that $g \in L_2^B$, $f \in L_1^A$ and the equalities $g^{\uparrow_\sigma} = f$ and $f^{\downarrow^\sigma} = g$ are satisfied.

Considering the theory of FCA, we want to provide a new mechanism to make recommendations related to the interest of one user. In the following section, we will introduce the data set we have considered in order to create our recommendation system.

3 Locate-Base Social Network

As we previously mentioned, in this work we will consider the information collected by a locate-base social network (LBSN). Therefore, we need to formalize this notion now.

Definition 4 ([10]) The LBSN (G, C) consists of a social network $G = (U, E)$, where U is the set of users, $E \subseteq U \times U$ is a relation between users, where $(u_i, u_j) \in E$ represents a social connection between two different users u_i and u_j ($u_i \neq u_j$); and $C \subseteq U \times L \times T$ is a relation in which each element $(u, l, t) \in C$ represents the location $l \in L$ at time $t \in T$ of one user $u \in U$, where L is a set of locations, which consists of latitude and longitude, and T is the set of times. Every element of C is called *check-in*.

Usually, the user communicates by some app his/her feeling, emotion, experience, etc., in some place, the information about this place, time and user (Id) is registered by the app and it is introduced as a check-in in the LBSN. Hence, every day the user can collect a lot of places in which his/her was interesting on. This information may represent the user's interest and habits. Since the user's check-in depends on his/her lifestyle and previous experiences, the check-ins also show the user's habits. Consequently, the study of the LBSN allows the analysis of the human's behaviors and customs which may be useful for themself and for business industry, marketing, advertisement, commerce, etc.

In LBSN theory, the check-in is completed with the characteristics of the location/place in which the user is. These characteristics are called *topics*. The association of each check-in with a topic is called *Point-of-Interest* (POI). Therefore, the set of POIs is a relation POI $\subseteq U \times L \times T \times P$, where P is the set of topics.

In this paper we have chosen a data set from the Foursquare social network. Table 1 shows a small example of the data set with four POI.

Concerning the topic column, sometimes the topic does not have a clear characteristic of the place in which the check-in has been produced. For example, the check-in for user 1 has two different characteristics: "nightlife" and "food", that do not exactly describe the place and so, the interest of the user. In order to determinate the user's interest, which we want to recommend him in one fixed time, we need to consider more check-ins at a similar hour.

Table 1 Example of the Foursquare data set

UserID	Data	Time	Loc(x, y)	Topic
1	29.07.2011	00:34:05	40.760997, −73.98290	Nightlife & Spot, Food
1891	21.07.2011	17:07:59	33.91865, −118.39331	Professional & Other Places
5	24.07.2010	00:20:26	47.60635, −122.33202	Nightlife & Spot, Food
882	31.01.2011	02:14:05	37.75026, −122.20290	Stadium, Arts & Entertainment
882	21.01.2011	19:51:20	37.79540, −122.3957	Food
884	06.08.2011	00.26.50	33.50887, −112.08372	Nightlife & Spot

This study will provide the usual interests of the user and the habits (s)he has and, as a consequence, we can offer him/her recommendations of places related to the user's interest and the current location (s)he has, such as, restaurants, theaters, sport centers, etc., near to the location in which (s)he is doing a new check-in.

4 Computing a Fuzzy Formal Context from the LBSM Dataset

In this section, we will show the considered procedure in order to obtain a fuzzy formal context and define a methodology for recommendation based on FCA. Note that we will study the interests and habits of only one user.

The procedure consists of three basic steps:

- Compute a partition of the check-ins by similar POI for the target user, comparing the time and topic of the check-ins.
- For each hour h and characteristic r (included in the topics), we compute a truth value of the sentence: "the user u at hour h is in a place with characteristic r". In this case, we need to take into account that each topic can have more than one characteristic. This is important since, as we previously commented, the topic might not exactly describe the really interest of the target user at that time. In order to cover this problem we have differentiate two cases: (1) if only one characteristic r has the topic at hour h, the POI is clear and so, this check-in is considered with truth value 1; (2) if the topic has different characteristics $r_1, \ldots, r_n$, at hour h, we consider as truth value for each characteristic 0.8. We have assuming this value due to a smaller value penalize so much this check-in. Finally, for each hour h and characteristic r we sum all the truth values obtained for each check-in.
- Normalize the obtained matrix in order to obtain a [0, 1]-fuzzy relation. Note that the matrix that we have obtained from the second step may not be a [0, 1]-fuzzy relation, which is needed in order to be considered as a relation in the fuzzy FCA framework. Hence, we need to normalize the values of the matrix. Since, we would

like to provide recommendations we need to carry out a global normalization, considering the maximum element in the whole matrix.

Hence, after applying the proposed procedure to the LBSN database, we obtain a [0, 1]-fuzzy relation R between hours and characteristics which can be associated with a formal context in which the objects are the hours, the attributes are the characteristics and the relation is R. From this context the fuzzy FCA theory can be applied.

5 Applying FCA for Analyzing the Database

In this section, we will apply this procedure to a user which has 456 check-ins in the considered LBSN database, from 26.04.2010 to 12.08.2011. It is natural to assume that the user was not online always. As we previously commented, the set of objects will be formed by the hours and the attributes will be the different characteristics:

$A = \{$Stadium, Arts & Entertainment, Professional & Other Places, Outdoors
Recreation, Beach, Music Venue, Food, Travel & Transport, Hotel, Spanish
Restaurant, Medical Center, Movie, Theater, Nightlife & Spot, Shop Service,
Residence, Office, Gym Fitness and etc.$\}$.

$B = \{00, 01, \ldots, 23\}$.

The next step is to compute the relation R between the objects and attributes. First of all, we apply the first step, and we obtain 207 partitions of the check-ins. Each partition represents the different topics at one hour h with topic p. For example, the partition associated with the topic: "Professional & Other Places" at 15:00h (until 16:00h) has 9 check-ins, which correspond to the numbers 24, 31, 32, 54, 124, 125, 128, 157 and 435, of the list of 456 check-ins of the target user. The second step computes the intermediate matrix. Table 2 represents one example of the obtained intermediate matrix, in which we only consider four hours and five characteristics.

Table 2 Example of the intermediate matrix

	Stadium	Arts & entertainment	Professional & O.P.	Food	Nightlife spot
13	0	0	0	0	0
14	0	0.8	0.8	0	0
15	0	0	0.8	0	0.8
16	0	0	6.8	0	0.8
17	0	0.8	11.4	0.8	0.8
18	0	1	10.4	0	0.8

Table 3 Normalization of Table 2

	Stadium	Arts & entertainment	Professional & O.P.	Food	Nightlife spot
13	0	0	0	0	0
14	0	0.0412	0.0412	0	0
15	0	0	0.0412	0	0.0412
16	0	0	0.3505	0	0.0412
17	0	0.0412	0.5876	0.0412	0.0412
18	0	0.0515	0.5361	0	0.0412

Table 4 Characteristic ranking by period

Characteristic	$(g_m)^{\uparrow}$	$(g_a)^{\uparrow}$	$(g_e)^{\uparrow}$
Professional & O. P.	0	0.5876	0.2887
Arts & Entertainmant	0	0.0412	0.1753
Outdoors recreation	0	0	0.134
Nightlife & Spot	0.0412	0.0412	0.0825
Stadium	0	0	0.0412
Food	0.0825	0	0.0412

Note that this matrix is not a [0, 1]-fuzzy relation, which is needed in order to be considered in a formal context. Hence, we need to normalize the values of the matrix, this normalization will be global dividing the elements in the matrix by the greatest one, obtaining the relation R. For example, from Table 2 we obtain the values in Table 3. Note that this table only shows a subrelation of R with 4 hours of the 24 hours and 5 characteristics of the total of 33.

After defining the fuzzy formal context (A, B, R), we can use the concept-forming operators in order to compute the concepts, where the frame will be given by the unit interval and the Gödel pair. Specifically, for the recommendation process we will compute the concepts which represent some particular period, such as the morning, afternoon and evening, which can be associated with the following fuzzy subset of the object set: $g_m: B \to [0, 1]$, $g_a: B \to [0, 1]$ and $g_e: B \to [0, 1]$, respectively, which are defined as follows:

$$
g_m(b) = \begin{cases} 0.5 & \text{if } b = 7 \\ 1 & \text{if } b = 8 \\ 0 & \text{otherwise} \end{cases} \quad
g_a(b) = \begin{cases} 1 & \text{if } b = 15 \\ 0.5 & \text{if } b = 16 \\ 0 & \text{otherwise} \end{cases} \quad
g_e(b) = \begin{cases} 1 & \text{if } b = 22 \\ 0.5 & \text{if } b = 23 \\ 0 & \text{otherwise} \end{cases}
$$

Therefore, in order to know the lifestyle of the target user in that moments, we compute the mapping $(g_m)^{\uparrow}$, $(g_a)^{\uparrow}$ and $(g_e)^{\uparrow}$. The Table 4 shows the characteristics for each period with the greatest truth values.

At the morning the user prefers "Food" places. This is a usual time for breakfast, which can be at home or maybe in some restaurant or coffee house. At the afternoon, between 3 and 4 p.m., the category "Professional & O.P." has the greatest truth value and the second one has a very small value. Therefore, we can suppose that the user is usually working at the afternoon. Sometimes he is also traveling or doing some cultural activities, but these are unusual occupations.

At the evening, we also have the user is working and we can suppose that this person is working at night or freelance. In this case, the category "Arts & Entertainment" has a large truth value then we can suppose that the user mades some creative work or when (s)he is not working (s)he likes attending cultural performances.

Therefore, using FCA we can describe the user habits and lifestyles and so, we can recommend the better places for him/her in the moment in which (s)he is using the application. For example, at the evening the app will recommend the best cultural performances in nearby places.

6   Conclusions and Future Work

In this work, we have presented a procedure to obtain formal contexts from LBSN databases. By means of this procedure, the theory of fuzzy FCA can be applied to extract information of the considered database. This information let us study the interests and habits of one user and make recommendations related to his interest. In the future we will apply more methodologies and properties in order to extract more information from the LBSN.

References

1. Belohlávek, R., Vychodil, V.: Attribute dependencies for data with grades II. Int. J. Gen. Syst. **46**(1), 66–92 (2017)
2. Cordero, P., Enciso, M., Mora, A., Ojeda-Aciego, M., Rossi, C.: Knowledge discovery in social networks by using a logic-based treatment of implications. Knowl. Based Syst. **87**, 16–25 (2015)
3. Cornejo, M.E., Medina, J., Ramírez-Poussa, E.: A comparative study of adjoint triples. Fuzzy Sets Syst. **211**, 1–14 (2013)
4. Ganter, B., Wille, R.: Formal Concept Analysis: Mathematical Foundation. Springer (1999)
5. Konecny, J., Krupka, M.: Block relations in formal fuzzy concept analysis. Int. J. Approx. Reason. **73**, 27–55 (2016)
6. Medina, J., Ojeda-Aciego, M.: Multi-adjoint t-concept lattices. Inf. Sci. **180**(5), 712–725 (2010)
7. Medina, J., Ojeda-Aciego, M., Ruiz-Calviño, J.: Formal concept analysis via multi-adjoint concept lattices. Fuzzy Sets Syst. **160**(2), 130–144 (2009)
8. Rodríguez-Jiménez, J.M., Cordero, P., Enciso, M., Mora, A.: Data mining algorithms to compute mixed concepts with negative attributes: an application to breast cancer data analysis. Math. Methods Appl. Sci., n/a–n/a (2016)
9. Shao, M.-W., Li, K.-W.: Attribute reduction in generalized one-sided formal contexts. Inf. Sci. **378**, 317–327 (2017)

10. Zhu, W.-Y., Peng, W.-C., Chen, L.-J., Zheng, K., Zhou, X.: Modeling user mobility for location promotion in location-based social networks. In: Proceedings of the 21th ACM SIGKDD International Conference on Knowledge Discovery and Data Mining, KDD'15, pp. 1573–1582. ACM, New York, NY, USA (2015)

Generalized Boolean Algebras and Applications

O. S. A. Bleblou, B. Šešelja and A. Tepavčević

Abstract A new notion of a lattice valued Boolean algebra is introduced. It is based on an algebra with two binary, a unary and two nullary operations, which is not a crisp Boolean algebra in general. The classical equality is replaced by a lattice valued equivalence so that the Boolean algebra identities are correspondingly satisfied. Main properties of the new introduced notion are proved, and a connection with the notion of a generalized lattice valued lattice is provided. As an application, the paper contains basic structures for developing generalized Boolean functions.

1 Introduction and Preliminaries

Boolean algebras and functions are main algebraic tools on which the information and communication technologies are based. For lattice valued (fuzzy) structures let us mention fuzzy groups, fuzzy lattices [1, 14], fuzzy Boolean algebras [4, 12, 13] and generally fuzzy algebras [3, 6]. A fuzzy equality is introduced by Höhle [8], and then used in [2, 3, 5]. Lattice valued identities were defined and investigated in [10]. The aim of this paper is to develop algebras which are less strict than a Boolean algebra, in order to investigate functions over such generalized structures. Then it would be possible to apply these algebras and functions to a wider class of problems arising in the fields of information and communication technology. In the present paper Ω-valued Boolean algebras are introduced, where Ω is a complete lattice of membership values, and E is a lattice-valued equality which replaces the classical one. Our approach and the notion of the Ω-valued Boolean algebra by the definitions and properties differs from other known generalizations of the Boolean algebra in the fuzzy setting.

Research supported by Ministry of Education, Science and Technological Development, Republic of Serbia, Grant No. 174013.

O. S. A. Bleblou · B. Šešelja · A. Tepavčević (✉)
Faculty of Sciences, Department of Mathematics and Informatics, University of Novi Sad, Novi Sad, Serbia
e-mail: andreja@dmi.uns.ac.rs

© Springer Nature Switzerland AG 2019
M. E. Cornejo et al. (eds.), *Trends in Mathematics and Computational Intelligence*, Studies in Computational Intelligence 796,
https://doi.org/10.1007/978-3-030-00485-9_16

139

A Boolean algebra is an algebra $(B, \sqcap, \sqcup, ', O, I)$ with two binary, one unary and two nullary operations, satisfying: both binary operations are commutative, distributive in both directions, unary operation $'$ has the properties of the complement, O is a neutral element for the operation $\sqcup$, I is a neutral element for the operation $\sqcap$. A Boolean algebra is a Boolean lattice, which means that $(B, \sqcap, \sqcup)$ is a lattice fulfilling distributive laws of binary operations, that O and I are the bottom and the top element respectively, and a' is the unique complement of a, for every $a \in B$. Moreover, an ordering relation $\leqslant$ can be defined in the Boolean algebra: $x \leqslant y$ iff $x \sqcap y = x$.

The membership values structure here is a complete lattice; it is a structure $(\Omega, \wedge, \vee, \leqslant)$ with ordering relation $\leqslant$, having the infimum and the supremum for every subset. Infimum and supremum of an arbitrary family $\{p_i \mid i \in I\}$ of elements from Ω are denoted by $\bigwedge_{i \in I} p_i$ and $\bigvee_{i \in I} p_i$, respectively. A complete lattice has the greatest element, 1, and the smallest element, 0.

A **lattice valued set** μ on a nonempty set A (or a **fuzzy set** on A) is a function $\mu : A \rightarrow \Omega$, where Ω is a complete lattice. A **cut set** of a lattice valued set $\mu : A \rightarrow \Omega$ is a subset μ_p of A defined by $\mu_p = \{x \in X \mid \mu(x) \geqslant p\}$.

Importance of so called "cutworthy" approach (the approach of dealing with cuts) is highlighted in [9].

A **lattice valued relation** R **on** A is a mapping $R : A^2 \rightarrow \Omega$. Let $\mu : A \rightarrow \Omega$ be an Ω-valued set on A and let $R : A^2 \rightarrow \Omega$ be an Ω-valued relation on A. If for all $x, y \in A$, R satisfies $R(x, y) \leqslant \mu(x) \wedge \mu(y)$, then we say that R is an Ω-**valued relation on** μ.

$$R \text{ is } \textbf{reflexive} \text{ on } \mu \text{ if } R(x, x) = \mu(x) \text{ for every } x \in A. \tag{1}$$

$$R \text{ is } \textbf{symmetric} \text{ if } R(x, y) = R(y, x) \text{ for all } x, y \in A; \tag{2}$$

$$R \text{ is } \textbf{transitive} \text{ if } R(x, z) \wedge R(z, y) \leqslant R(x, y) \text{ for all } x, y, z \in A. \tag{3}$$

A reflexive, symmetric and transitive relation R on μ is a **lattice valued equivalence** on μ.

If E is a lattice valued equivalence on a lattice valued set μ, $\mu(x) = E(x, x)$, we say that (A, E) is an Ω-**set**.

Clearly, for $p \in \Omega$, the cut E_p of E is an equivalence relations on μ_p.

Let E be an Ω-valued equivalence on μ. An Ω-valued relation $R : A^2 \rightarrow \Omega$ on A is E-**antisymmetric**, if the following holds:

$$R(x, y) \wedge R(y, x) = E(x, y), \quad \text{for all } x, y \in A. \tag{4}$$

Let (A, E) be an Ω-set. An Ω-valued relation $R : A^2 \rightarrow \Omega$ on A is an Ω-**valued order** on (A, E), if R is reflexive on μ, E-antisymmetric, and transitive.

Let $\mathcal{A} = (A, F)$ be an algebra, where A is a nonempty set, and F is a set of operations on A. A **lattice valued subalgebra** of $\mathcal{A}$ is any mapping $\mu : A \to \Omega$, which fulfils the following: For any operation f from F with an arity $n > 0$, $f : A^n \to A$, and for all $a_1, \ldots, a_n \in A$, the following is satisfied

$$\bigwedge_{i=1}^{n} \mu(a_i) \leqslant \mu(f(a_1, \ldots, a_n)), \tag{5}$$

and for a nullary operation (constant) $c \in F$, $\mu(c) = 1$.

Let $A = (A, F)$ be an algebra. A fuzzy relation $R : A^2 \to \Omega$ is **compatible** with the operations in F if the following holds: for every n-ary operation $f \in F$ and for all $a_1, \ldots, a_n, b_1, \ldots, b_n \in A$

$$\bigwedge_{i=1}^{n} R(a_i, b_i) \leqslant R(f(a_1, \ldots, a_n), f(b_1, \ldots, b_n)), \text{ and} \tag{6}$$

$$R(c, c) = 1 \text{ for every constant (nullary operation) } c \in F. \tag{7}$$

If R is an Ω-valued relation on an Ω-valued subalgebra μ of $\mathcal{A}$, then it is **compatible** (or compatible on μ), if it is compatible with the operations in F, in the sense of (6) and (7).

A **separated** Ω-valued equivalence on an Ω-valued subalgebra μ is a compatible Ω-valued equivalence on μ, fulfilling (see e.g., [10]): for all $x, y \in A$, $x \neq y$, if $R(x, x) \neq 0$, then $R(x, y) < R(x, x)$. A separated Ω-valued equivalence is in the fuzzy framework frequently called an Ω-**valued equality**, since the classical equality which is replaced by an Ω-valued equivalence, is also separated.

Let $u(x_1, \ldots, x_n) \approx v(x_1, \ldots, x_n)$ be an identity in the language of an algebra $\mathcal{A}$. Let also μ be an Ω-valued subalgebra of $\mathcal{A}$, and E an Ω-valued equivalence on $\mathcal{A}$. We say that the above identity **holds (is valid)** on μ **with respect to** E if the following condition is fulfilled for all $a_1, \ldots, a_n \in A$:

$$\bigwedge_{i=1}^{n} \mu(a_i) \leqslant E(u(a_1, \ldots, a_n), v(a_1, \ldots, a_n)). \tag{8}$$

If an Ω-valued subalgebra μ of an algebra $\mathcal{A}$ fulfils an identity $u \approx v$, then this identity need not hold on $\mathcal{A}$. However, the converse holds, i.e., an identity fulfilled by the basic algebra is also satisfied by the corresponding Ω-valued subalgebra μ with respect to E.

In papers [7, 11] a new concept of fuzzy lattices has been developed as an Ω-structure. The basic structure is an algebra with two binary operations $(M, \sqcap, \sqcup)$, and an Ω-valued equivalence $E : M^2 \to \Omega$, compatible with operations $\sqcap$ and $\sqcup$. Then $\mathcal{M} = (M, E)$ is an Ω-**valued lattice** if the known lattice identities hold. By (8), this means that the following formulas are satisfied:

$$\mu(x) \wedge \mu(y) \leqslant E(x \sqcap y, y \sqcap x)$$
$$\mu(x) \wedge \mu(y) \leqslant E(x \sqcup y, y \sqcup x)$$
$$\mu(x) \wedge \mu(y) \wedge \mu(z) \leqslant E((x \sqcap y) \sqcap z, x \sqcap (y \sqcap z))$$
$$\mu(x) \wedge \mu(y) \wedge \mu(z) \leqslant E((x \sqcup y) \sqcup z, x \sqcup (y \sqcup z))$$
$$\mu(x) \wedge \mu(y) \leqslant E((x \sqcap y) \sqcup x, x)$$
$$\mu(x) \wedge \mu(y) \leqslant E((x \sqcup y) \sqcap x, x).$$

2 Ω-valued Boolean Algebra: Definition and Results

Let $\mathcal{B} = (B, \sqcap, \sqcup, ', O, I)$ be an algebraic structure with two binary, one unary and two nullary operations (constants) and let Ω be a complete lattice with the top and the bottom element 1 and 0 respectively. Let (B, E) be an Ω-set, where E is compatible with all the operations of $\mathcal{B}$.

The ordered pair $(\mathcal{B}, E)$ is an Ω-**valued Boolean algebra** if the classical axioms for Boolean algebras are fulfilled. This means:

1. $\mu(x) \wedge \mu(y) \leq E(x \sqcap y, y \sqcap x)$
2. $\mu(x) \wedge \mu(y) \leq E(x \sqcup y, y \sqcup x)$
3. $\mu(x) \wedge \mu(y) \wedge \mu(z) \leq E(x \sqcap (y \sqcup z), (x \sqcap y) \sqcup (x \sqcap z))$
4. $\mu(x) \wedge \mu(y) \wedge \mu(z) \leq E(x \sqcup (y \sqcap z), (x \sqcup y) \sqcap (x \sqcup z))$
5. $\mu(x) \leq E(x \sqcup O, x)$
6. $\mu(x) \leq E(x \sqcap I, x)$
7. $\mu(x) \leq E(x \sqcap x', O)$
8. $\mu(x) \leq E(x \sqcup x', I)$
9. $E(O, I) < 1$.

Straightforwardly, $\mu(x) = E(x \sqcup O, x)$ and similarly $\mu(x) = E(x \sqcap I, x)$. In case E is separated, we obtain $x \sqcup O = x$ and $x \sqcap I = x$ meaning that in this case $\mathcal{B}$ should have neutral elements for $\sqcup$ and $\sqcap$.

Since the formulas corresponding to axioms are dual in the sense that they appear in the dual pairs w.r.t. $\sqcap$ and $\sqcup$, also O and I, **the principle of duality** is satisfied. This means that for every statement which is true in the language of algebra $(B, \sqcap, \sqcup, ', O, I)$, the dual statement is also true. The dual statement is obtained exchanging each occurrence of $\sqcap$ with $\sqcup$ and vice versa and exchanging each occurrence of O with I and vice versa.

Proposition 1 *Let $\mathcal{B} = (B, \sqcap, \sqcup, ', O, I)$ be a Boolean algebra, Ω a complete lattice and let $\mu : B \to \Omega$ be a lattice valued subalgebra of $\mathcal{B}$. If E is an arbitrary Ω-valued equivalence on μ, then, $(\mathcal{B}, E)$ is an Ω-Boolean algebra.*

Proposition 2 *Let $\mathcal{B} = (B, \sqcap, \sqcup, ', O, I)$ be an algebraic structure, Ω a complete lattice, and $(\mathcal{B}, E)$ an Ω-Boolean algebra. Then, the following identities hold on $(\mathcal{B}, E)$:*

$$x \sqcap O \approx O \quad and \quad x \sqcup I \approx I;$$
$$x \sqcap (x \sqcup y) \approx x \quad and \quad x \sqcup (x \sqcap y) \approx x;$$
$$x \sqcap x \approx x \quad and \quad x \sqcup x \approx x;$$
$$x \sqcap (y \sqcap z) \approx (x \sqcap y) \sqcap z; \quad and \quad x \sqcap (y \sqcap z) \approx (x \sqcap y) \sqcap z.$$

Ω-Boolean algebra is also an Ω-lattice, as follows.

Theorem 1 *Let* $(B, \sqcap, \sqcup, ', O, I)$ *be an algebraic structure as above,* Ω *a complete lattice, and* (B, E) *an* Ω-*Boolean algebra. Then* (M, E) *is an* Ω-*lattice, where* $M = (B, \sqcap, \sqcup)$ *is a bi-groupoid which is a reduct of the starting structure.*

If $(B, \sqcap, \sqcup, ', O, I)$ is an algebraic structure and $\mu : B \to \Omega$ a lattice valued algebra on μ, then for every $p \in \Omega$, μ_p are subalgebras of $(B, \sqcap, \sqcup, ', O, I)$. Moreover, if $E : B^2 \to \Omega$ is an Ω-valued equivalence on μ, then all the cut relations E_p, are congruences on μ_p for $p \in \Omega$.

Theorem 2 *Let* $\mathcal{B} = (B, \sqcap, \sqcup, ', O, I)$ *be an algebraic structure with two binary operation, one unary and two constants, and* Ω *a complete lattice. Then,* $(\mathcal{B}, E)$ *is an* Ω-*Boolean algebra if and only if for every* $p \in \Omega$, *the quotient structure* μ_p/E_p *is a (classical) Boolean algebra.*

Let $(\mathcal{B}, E)$ be an Ω-Boolean algebra with $(B, \sqcap, \sqcup, ', O, I)$ being an algebraic structure as above, and $\mu(x) = E(x, x)$.

Proposition 3 *Let* $(\mathcal{B}, E)$ *be an* Ω-*Boolean algebra, with* $(B, \sqcap, \sqcup, ', O, I)$ *being an algebraic structure with two binary, one unary and two nullary operations and* E *a separated* Ω-*valued equivalence on* $\mathcal{B}$. *Then an* Ω-*valued relation* $R : B^2 \to \Omega$, *defined by* $R(x, y) := \mu(x) \wedge \mu(y) \wedge E(x \sqcap y, x)$ *is an* Ω-*valued order on* $(\mathcal{B}, E)$.

Our main interest in Ω-Boolean algebras are those in which the basic structure is a collection of n-tuples over the two-element set $\{0, 1\}$. In other words, we are mostly concentrated to Ω-Boolean algebras $(\mathcal{B}, E)$, where

$$\mathcal{B} = (B, \sqcap, \sqcup, \bar{\ }, O, I), \quad B \subseteq \{0, 1\}^n, \tag{9}$$

while the operations $\sqcap$, $\sqcup$, and $\bar{\ }$ are arbitrary (two binary and a unary one, respectively), $O = (0, 0, \ldots, 0)$, $I = (1, 1, \ldots, 1)$. As usual, we denote by μ the function $\mu : B \to \Omega$, such that for every n-tuple $x \in B$, we have $\mu(x) = E(x, x)$.

These finite sequences of zeros and ones are codewords in the digital technology and the above structure is usually complete (consisting of the whole set $\{0, 1\}^n$), moreover it is a classical Boolean algebra. However, in reality noise and errors have an impact to the operations, and the Boolean structure may be corrupted to some extent; in addition, some tuples might be missing. The above Ω-Boolean algebra with suitable operations and with a fuzzy equality could be a model of such a modified structure.

Let us denote the classical Boolean algebra of all n-tuples of 0 and 1 as follows:
$\mathcal{B}_2^n = (\{0, 1\}^n, \min, \max, ', 0, 1)$, where, as usual, operations are defined componen-twise: for $(a_1, \ldots, a_n), (b_1, \ldots, b_n) \in \{0, 1\}^n$,

$$\min((a_1, \ldots, a_n), (b_1, \ldots, b_n)) = (\min(a_1, b_1), \ldots, \min(a_n, b_n));$$
$$\max((a_1, \ldots, a_n), (b_1, \ldots, b_n)) = (\max(a_1, b_1), \ldots, \max(a_n, b_n));$$
$$(a_1, \ldots, a_n)' = (a_1', \ldots, a_n').$$

We say that an Ω-Boolean algebra is **standard** if it is of the form $(\mathcal{B}_2^n, H)$, $H : (\{0, 1\}^n)^2 \to \Omega$ being an Ω-valued equivalence.

In the sequel, we deal with Ω-Boolean algebras of the type (9), namely those in which $B \subseteq \{0, 1\}^n$, for some natural number n. Let us call such an Ω-Boolean algebra $((B, \sqcap, \sqcup, \bar{\ }, O, I), E), B \subseteq \{0, 1\}^n$ **regular**, if there is a standard Ω-Boolean algebra $(\mathcal{B}_2^n, H)$, such that the following hold:

(i) The Ω-valued equivalence $E : B \to \Omega$ is a restriction of the Ω-valued equivalence H, i.e., $E = H|_B$.

(ii) For all n-tuples $x, y \in B$,

    (a) $E(x, x) \leqslant E(x', \bar{x})$;
    (b) $E(x, x) \wedge E(y, y) \leqslant E(x \sqcap y, \min(x, y))$;
    (c) $E(x, x) \wedge E(y, y) \leqslant E(x \sqcup y, \max(x, y))$.

Theorem 3 *For a regular Ω-Boolean algebra the following holds:*

(i) For every $p \in \Omega$, $E_p \subseteq \Theta$, where $E_p = E^{-1}(\uparrow p)$ is the p-cut of the Ω-valued equivalence E, and Θ is a congruence on a Boolean subalgebra $\mathcal{M}$ of $\{0, 1\}^n$.

(ii) For every $p \in \Omega$, the map $[a]_{E_p} \mapsto [a]_\Theta$ is an isomorphism of the quotient Boolean algebra μ_p/E_p onto the Boolean algebra $\mathcal{M}/\Theta$, with notation as in (i).

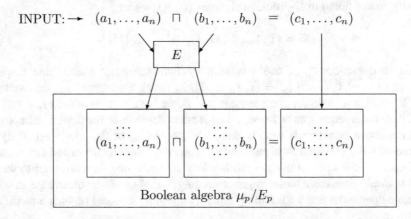

Fig. 1 Procedure by which computations are performed in a regular Ω-Boolean algebra

The procedure by which the computations are performed in a regular Ω-Boolean algebra is presented in Fig. 1.

3 Conclusion

In this paper we introduce an Ω-Boolean algebra. The domain structure need not be a Boolean algebra. Still, there is a connection: particular quotients of cuts of the lattice valued domain over the cuts of the corresponding lattice valued equality are Boolean algebras. Our intention is to use Ω-Boolean functions for investigations in generalizing switching and logical circuits and their applications.

References

1. Ajmal, N., Thomas, K.V.: Fuzzy lattices. Inf. Sci. **79**, 271–291 (1994)
2. Bělohlávek, R.: Fuzzy Relational Systems: Foundations and Principles. Kluwer Academic/Plenum Publishers, New York (2002)
3. Bělohlávek, R., Vychodil, V.: Algebras with fuzzy equalities. Fuzzy Sets Syst. **157**, 161–201 (2006)
4. Coulon, J., Coulon, J.-L.: Fuzzy boolean algebras. J. Math. Anal. Appl. **99**, 248–256 (1984)
5. Demirci, M.: A theory of vague lattices based on many-valued equivalence relations I: general representation results. Fuzzy Sets Syst. **151**, 437–472 (2005)
6. Di Nola, A., Gerla, G.: Lattice valued algebras. Stochastica **11**, 137–150 (1987)
7. Edeghagba Eghosa, E., Šešelja, B., Tepavčević, A.: Ω-lattices. Fuzzy Sets Syst. **311**, 53–69 (2017)
8. Höhle, U.: Quotients with respect to similarity relations. Fuzzy Sets Syst. **27**, 31–44 (1988)
9. Klir, G., Yuan, B.: Fuzzy Sets and Fuzzy Logic. Prentice Hall PTR, New Jersey (1995)
10. Šešelja, B., Tepavčević, A.: Fuzzy identities. In: Proceedings of the 2009 IEEE International Conference on Fuzzy Systems, pp. 1660–1664
11. Šešelja, B., Tepavčević, A.: L-E-fuzzy lattices. Int. J. Fuzzy Syst. **17**(3), 366–374 (2015)
12. Šešelja, B., Tepavčević, A.: Fuzzy Boolean algebra. In: Proceedings of the IFIP TC12/WG12.3 International Workshop on Automated Reasoning, pp. 83–88 (1992)
13. Swamy, U.M., Murthy, M.K.: Representation of fuzzy Boolean algebras. Fuzzy Sets Syst. **48**, 231–237 (1992)
14. Tepavčević, A., Trajkovski, G.: L-fuzzy lattices: an introduction. Fuzzy Sets Syst. **123**, 209–216 (2001)

Bipolar Max-Product Fuzzy Relation Equations with the Product Negation

M. Eugenia Cornejo, David Lobo and Jesús Medina

Abstract This paper will study the bipolar fuzzy relation equation based on the max-product composition and the adjoint negation operator obtained from the product residuated implication. Interesting properties and different examples of this bipolar max-product fuzzy relation equation will be introduced.

1 Introduction

Fuzzy relation equations (FREs), introduced by Sanchez [17, 18], have been widely studied [1, 5–8, 10, 13–16] due to their application in approximate reasoning, decision making and fuzzy control, etc. Considering these fuzzy relation equations containing unknown variables together with their logical negations simultaneously give rise to a new kind of equations called bipolar fuzzy relation equations. These bipolar equations have already been used in applications associated with optimization problems, where the variables need to show a bipolar character [9, 11, 19].

There are few papers dealing with bipolar fuzzy relation equations and they are devoted to the study of bipolar max-min fuzzy relation equations. In [12], we can found an elaborated analysis on the solvability of bipolar max-min fuzzy relation equations. Following this research line, we have already provided definitions, illustrative examples and interesting properties related to the solvability of bipolar max-

---

Partially supported by the State Research Agency (AEI) and the European Regional Development Fund (ERDF) project TIN2016-76653-P, and by the research and transfer program of the University of Cádiz.

M. E. Cornejo (✉) · D. Lobo · J. Medina
Department of Mathematics, University of Cádiz, Cádiz, Spain
e-mail: mariaeugenia.cornejo@uca.es

D. Lobo
e-mail: david.lobo@uca.es

J. Medina
e-mail: jesus.medina@uca.es

© Springer Nature Switzerland AG 2019
M. E. Cornejo et al. (eds.), *Trends in Mathematics and Computational Intelligence*, Studies in Computational Intelligence 796,
https://doi.org/10.1007/978-3-030-00485-9_17

product fuzzy relation equations with the standard negation [2]. In this paper, we consider these equations with another negation, the residuated negation associated with the product t-norm, which is not involutive and provides different solvability properties as we will show in this paper. These studies will be fundamental in the development of future advances towards the consideration of arbitrary negations.

2 Solving Bipolar FREs Based on the Product T-Norm

The basic operators involved in the studied bipolar FREs will be the product adjoint pair $(\&_P, \leftarrow_P)$ defined as $x \&_P y = x * y, z \leftarrow_P x = \min(1, z/x)$, for all $x, y, z \in [0, 1]$, and the product adjoint negation n_P defined as $n_P(0) = 1$ and $n_P(x) = 0$, for all $x \in]0, 1]$, which will be simply called *product negation*. Note that this negation is also obtained from the residuated negation of others left-continuous t-norms, such as, the Gödel t-norm [3, 4].

In this paper, first of all, we will introduce the notion of bipolar max-product FRE with the product negation and a unique unknown variable. Then, we will analyze simple scenarios and we will provide a characterization on the solvability of these equations. The next step in our paper will consist in giving sufficient and necessary conditions which allow us to ensure when bipolar max-product FREs with different unknown variables are solvable. In addition, different results related to the existence of the greatest (least, respectively) solution or a finite number of maximal (minimal, respectively) solutions for these last equations are presented.

2.1 A Simple Scenary

Our study will start with the most simple case of bipolar max-product FRE, that is, containing a unique unknown variable. Such equation is given as follows:

$$(a^+ * x) \vee (a^- * n_P(x)) = b \tag{1}$$

where $a^+, a^-, b \in [0, 1], x$ is an unknown variable in $[0, 1]$, $*$ is the product operator, $\vee$ is the maximum operator and n_P is the product adjoint negation.

It is important to note that Eq. (1) is simplified when either a^+, a^- or b is equal to 0, and as a consequence its solvability can be characterized in an easier way. In what follows, the previous three different cases will be considered in order to obtain a first approach on the solvability of Eq. (1).

(i) If $a^+ = 0$, then Eq. (1) is given by $a^- * n_P(x) = b$. According to the definition of n_P, we can deduce that Eq. (1) is solvable if $b = 0$ or $a^- = b$. When $b = 0$, the solution of Eq. (1) is any $x \in [0, 1[$. If $a^- = b$ then the unique solution of Eq. (1) is $x = 0$.

(ii) If $a^- = 0$, then Eq. (1) becomes into $a^+ * x = b$. Taking into account that the product t-norm is a continuous order-preserving mapping and considering that the equalities $a^+ * 0 = 0$ and $a^+ * 1 = a^+$ hold, we can ensure that Eq. (1) is solvable if and only if $a^+ \geq b$.

(iii) If $b = 0$, then Eq. (1) is given by $(a^+ * x) \vee (a^- * n_P(x)) = 0$. This last equation is solvable if and only if $a^+ * x = 0$ and $a^- * n_P(x) = 0$. From the definition of the negation n_P, we can conclude that Eq. (1) is solvable if and only if $a^+ = 0$ or $a^- = 0$. Note that these cases have already been studied above.

From now on, we will consider that each known variable appearing in bipolar max-product t-norm FREs is different from zero. The following result characterizes the solvability of bipolar max-product FREs defined with the adjoint negation associated with the product t-norm.

Theorem 1 *Let $a^+, a^-, b \in (0, 1]$ and x an unknown variable belonging to $[0, 1]$. The bipolar max-product t-norm FRE given by Eq. (1) is solvable if and only if $a^- = b$ or $a^+ \geq b$.*

Notice that, Eq. (1) has at most two solutions when it is solvable. These solutions are $x = 0$ and $x = b \leftarrow_P a^+$, being $\leftarrow_P$ the residuated implication related to the product t-norm. When both $x = 0$ and $x = b \leftarrow_P a^+$ are solutions, then $x = b \leftarrow_P a^+$ is the greatest one. We will illustrate the above theorem by means of the following example.

Example 1 Given the bipolar max-product t-norm FRE defined as

$$(0.5 * x) \vee (0.2 * n_P(x)) = 0.3 \tag{2}$$

Equation (2) is solvable since the hypothesis required in Theorem 1 are satisfied. Specifically, we have that $0.2 \neq 0.3$ and $0.5 \geq 0.3$, from which we can ensure that $x = 0.3 \leftarrow_P 0.5 = 0.6$ is the unique solution of Eq. (2):

$$(0.5 * 0.6) \vee (0.2 * n_P(0.6)) = (0.5 * 0.6) \vee (0.2 * 0) = 0.3 \vee 0 = 0.3$$

It is easy to proved that $x = 0$ is not a solution of Eq. (2). If we replace the variables we have: $(0.5 * 0) \vee (0.2 * n_P(0)) = (0.5 * 0) \vee (0.2 * 1) = 0 \vee 0.2 = 0.2$, which is different from 0.3   □

Now, we are interested in studying more complex bipolar max-product fuzzy relation equations. We will continue our research with a bipolar max-product fuzzy relation equation containing different unknown variables.

2.2 A Bipolar Max-Product FRE with Different Unknown Variables

In this section, we will provide sufficient and necessary conditions to guarantee the solvability of bipolar max-product FREs with the product negation, in which appear a finite number of different unknown variables. Moreover, we will show when these equations has either greatest (least, respectively) solution or a finite number of maximal (minimal, respectively) solutions. To begin with, we will show under what conditions a bipolar max-product FRE containing different unknown variables is solvable.

Theorem 2 *Let $a_i^+, a_i^-, b \in (0, 1]$ and x_i an unknown variable belonging to $[0, 1]$, for all $i \in \{1, \ldots, m\}$. The bipolar max-product fuzzy relation equation*

$$\bigvee_{i=1}^{m} (a_i^+ * x_i) \vee (a_i^- * n_P(x_i)) = b \tag{3}$$

is solvable if and only if $\max\{a_i^+ \mid i \in \{1, \ldots, m\}\} \geq b$ or there exists an index $k \in \{1, \ldots, m\}$ such that $a_k^- = b$.

Example 2 Consider the bipolar max-product FRE with three unknown variables $x_1, x_2, x_3 \in [0, 1]$ given by Eq. (4):

$$(0.4 * x_1) \vee (0.7 * n_P(x_1)) \vee (0.2 * x_2) \vee (0.1 * n_P(x_2)) \vee (0.5 * x_3) \vee (0.2 * n_P(x_3)) = 0.3 \tag{4}$$

Since $\max\{0.4, 0.2, 0.5\} = 0.5 \geq 0.3$, we can apply Theorem 2 what allows us to ensure that Eq. (4) is solvable. For instance, the tuple $(0.5, 0, 0.6)$ is a solution of such equation. Replacing the variables in the equation we have:

$$(0.4 * 0.5) \vee (0.7 * n_P(0.5)) \vee (0.2 * 0) \vee (0.1 * n_P(0)) \vee (0.5 * 0.6) \vee (0.2 * n_P(0.6))$$

which is equal to $0.2 \vee 0 \vee 0 \vee 0.1 \vee 0.3 \vee 0 = 0.3$.   □

Although we have provided only one solution for Eq. (4) in the example above, we can make simple computations in order to obtain all its possible solutions. Moreover, one can wonder if there exists the greatest solution or several maximal solutions for Eq. (4). An analogous question arises with respect to the existence of its least solution or different minimal solutions. In order to clarify these issues for a general bipolar max-product FRE with different variables, we introduce two interesting results. These results use the notion of cardinality of one set A, which will be denoted as $\text{card}(A)$.

Theorem 3 *Given $a_i^+, a_i^-, b \in (0, 1]$ and x_i an unknown variable in $[0, 1]$, for all $i \in \{1, \ldots, m\}$. Consider a solvable bipolar max-product FRE*

$$\bigvee_{i=1}^{m} (a_i^+ * x_i) \vee (a_i^- * n_P(x_i)) = b \tag{5}$$

Then, the following statements hold:

(1) If $\max\{a_i^+ \mid i \in \{1, \ldots, m\}\} \geq b$, *the set of solutions of Eq. (5) has a greatest element.*

(2) If $a_i^+ < b$ *for all* $i \in \{1, \ldots, m\}$, *then the set of maximal solutions of Eq. (5) is finite. Moreover, the number of maximal solutions is equal to:*

$$card(\{k \in \{1, \ldots, m\} \mid a_k^- = b\})$$

A similar result is obtained with respect to the least solution and the set of minimal solutions of a solvable bipolar max-product FRE with different variables. Nevertheless, this result shows that the set of minimal solutions of such equation can be empty.

Theorem 4 *Given* $a_i^+, a_i^-, b \in (0, 1]$ *and* x_i *an unknown variable belonging to* $[0, 1]$, *for all* $i \in \{1, \ldots, m\}$. *Consider a solvable bipolar max-product FRE*

$$\bigvee_{i=1}^{m} (a_i^+ * x_i) \vee (a_i^- * n_P(x_i)) = b \tag{6}$$

Then, the following statements hold:

(1) If there exists $k \in \{1, \ldots, m\}$ *such that* $a_k^- = b$ *and* $a_i^- \leq b$, *for all* $i \in \{1, \ldots, m\}$, *then the set of solutions of Eq. (6) has a least element.*

(2) If there exist $k_1, k_2 \in \{1, \ldots, m\}$ *such that* $a_{k_1}^- = b$ *and* $a_{k_2}^- > b$, *then the set of solutions of Eq. (6) has no minimal elements.*

(3) If $a_i^- \neq b$ *for all* $i \in \{1, \ldots, m\}$, *then the set of minimal solutions of Eq. (6) is finite. Moreover, the number of minimal solutions is:*

$$card(\{k \in \{1, \ldots, m\} \mid a_k^+ \geq b \text{ and } a_i^- \leq b \text{ for all } i \neq k\})$$

In order to illustrate Theorems 3 and 4, we will carry on with the study of the bipolar max-product FRE given in Example 2.

Example 3 Coming back to Example 2, we have obtained that Eq. (4) is solvable and therefore, we can apply Theorems 3 and 4. Equation (4) has a greatest solution because statement (1) in Theorem 3 is verified, that is $\max\{0.4, 0.2, 0.5\} \geq 0.3$. The greatest solution is the tuple $(0.75, 1, 0.6)$.

In addition, it is easy to see that the conditions required in statements (3) in Theorem 4 are satisfied. According to statement (3), we obtain that the number of minimal solutions of Eq. (4) is:

$$card(\{k \in \{1, 2, 3\} \mid a_k^+ \geq b \text{ and } a_i^- \leq b \text{ for all } i \neq k\}) = card\{1\} = 1$$

As a consequence, we can deduce that Eq. (4) has only one minimal solution which is $(0.75, 0, 0)$. It is important to mention that $(0, 0, 0.6)$ is not a solution of Eq. (4), as you can check below:

$$(0.4 * 0) \vee (0.7 * 1) \vee (0.2 * 0) \vee (0.1 * 1) \vee (0.5 * 0.6) \vee (0.2 * 0) = 0.7$$

On the other hand, if we modify Eq. (4) obtaining the following one:

$$(0.4 * x_1) \vee (0.7 * n_P(x_1)) \vee (0.2 * x_2) \vee (0.1 * n_P(x_2)) \vee (0.5 * x_3) \vee (0.9 * n_P(x_3)) = 0.3 \tag{7}$$

We can apply Theorems 2 and 4 in order to assert that Eq. (7) is solvable but it has no minimal solutions. Specifically, applying Theorem 4, we have that the number of minimal solutions in Eq. (7) is:

$$\text{card}(\{k \in \{1, 2, 3\} \mid a_k^+ \geq b \text{ and } a_i^- \leq b \text{ for all } i \neq k\}) = \text{card}\{\varnothing\} = 0 \qquad \square$$

After presenting the characterization about the solvability of bipolar max-product FREs with the product negation and the conditions to guarantee the existence of their maximal and/or minimal solutions, we will focus on our next goal. We are interested in solving bipolar max-product FREs systems.

3   Conclusions and Future Work

This paper has been focused on the resolution of bipolar max-product FREs with the product negation. Although the obtained results have continued in the research line proposed in [2], the conditions required to guarantee the solvability of bipolar max-product FREs with different variables are different. Clearly, these differences are due to the use of the product negation (non-involutive negation) instead of the standard negation (involutive negation). Hence, we have shown the interest in the study of this bipolar FREs with the particular case of the product negation. Therefore, we will study in the future the solvability of systems of this kind of bipolar FREs with one or more unknown variables.

References

1. Bělohlávek, R.: Sup-t-norm and inf-residuum are one type of relational product: Unifying framework and consequences. Fuzzy Sets Syst. **197**, 45–58 (2012)
2. Cornejo, M., Lobo, D., Medina, J.: Bipolar fuzzy relation equations based on product t-norm. In: 2017 IEEE International Conference on Fuzzy Systems. IEEE Press (2017)
3. Cornejo, M.E., Medina, J., Ramírez-Poussa, E.: A comparative study of adjoint triples. Fuzzy Sets Syst. **211**, 1–14 (2013)
4. Cornejo, M.E., Medina, J., Ramírez-Poussa, E.: Adjoint negations, more than residuated negations. Inf. Sci. **345**, 355–371 (2016)
5. De Baets, B.: Analytical solution methods for fuzzy relation equations. In: Dubois, D., Prade, H. (eds.) The Handbooks of Fuzzy Sets Series, vol. 1, pp. 291–340. Kluwer, Dordrecht (1999)
6. Di Nola, A., Sanchez, E., Pedrycz, W., Sessa, S.: Fuzzy Relation Equations and Their Applications to Knowledge Engineering. Kluwer Academic Publishers, Norwell, MA, USA (1989)

7. Díaz, J.C., Medina, J.: Multi-adjoint relation equations: definition, properties and solutions using concept lattices. Inf. Sci. **253**, 100–109 (2013)
8. Díaz-Moreno, J.C., Medina, J., Turunen, E.: Minimal solutions of general fuzzy relation equations on linear carriers. an algebraic characterization. Fuzzy Sets Syst. **311**, 112–123 (2017)
9. Freson, S., Baets, B.D., Meyer, H.D.: Linear optimization with bipolar max-min constraints. Inf. Sci., **234**, 3–15 (2013). Fuzzy Relation Equations: New Trends and Applications
10. Ignjatović, J., Ćirić, M., Šešelja, B., Tepavčević, A.: Fuzzy relational inequalities and equations, fuzzy quasi-orders, closures and openings of fuzzy sets. Fuzzy Sets Syst. **260**, 1–24 (2015). Theme: Algebraic Structures
11. Li, D.-C., Geng, S.-L.: Optimal solution of multi-objective linear programming with inf-$\rightarrow$ fuzzy relation equations constraint. Inf. Sci. **271**, 159–178 (2014)
12. Li, P., Jin, Q.: On the resolution of bipolar max-min equations. Kybernetika **52**(4), 514–530 (2016)
13. Medina, J.: Minimal solutions of generalized fuzzy relational equations: Clarifications and corrections towards a more flexible setting. Int. J. Approx. Reason. **84**, 33–38 (2017)
14. Medina, J.: Notes on 'solution sets of inf-α_T fuzzy relational equations on complete brouwerian lattice' and 'fuzzy relational equations on complete brouwerian lattices'. Inf. Sci. **402**, 82–90 (2017)
15. Peeva, K.: Imprecision and Uncertainty in Information Representation and Processing: New Tools Based on Intuitionistic Fuzzy Sets and Generalized Nets, pp. 73–85. Springer International Publishing, Cham (2016)
16. Perfilieva, I., Nosková, L.: System of fuzzy relation equations with inf-$\rightarrow$ composition: complete set of solutions. Fuzzy Sets Syst. **159**(17), 2256–2271 (2008)
17. Sanchez, E.: Resolution of composite fuzzy relation equations. Inf. Control **30**(1), 38–48 (1976)
18. Sanchez, E.: Inverses of fuzzy relations. application to possibility distributions and medical diagnosis. Fuzzy Sets Syst. **2**(1), 75–86 (1979)
19. Zhou, J., Yu, Y., Liu, Y., Zhang, Y.: Solving nonlinear optimization problems with bipolar fuzzy relational equation constraints. J. Inequal. Appl. **2016**(1), 126 (2016)

Direct and Indirect Methods for Solving Two-Mode Systems of Fuzzy Relation Equations and Inequalities

Miroslav Ćirić, Jelena Ignjatović and Ivan Stanković

Abstract The purpose of this paper is to compare two methods for computing the greatest solutions of two-mode systems of FREIs. The first one is the direct method developed recently in [26], and the second one consists in converting the two-mode case into the one-mode case, and computing the greatest solutions of related one-mode systems of FREIs, using algorithms provided in [13, 15]. The conversion is made by means of the well-known method used in social network analysis for transforming two-mode networks into one-mode networks. We prove theoretically interesting result according to which solutions of any two-mode system of FREIs can be derived from solutions of the related one-mode system, and vice versa. However, from the computational point of view, the conversion based method is more memory and time demanding, what favors the direct method.

1 Introduction and Preliminaries

The study of systems of fuzzy relation equations and inequalities (FREIs) was initiated by E. Sanchez, who used them in medical research (cf. [21–23]). Later they found a much wider field of application, and nowadays they are used in fuzzy control, discrete dynamic systems, knowledge engineering, identification of fuzzy systems, prediction of fuzzy systems, decision-making, fuzzy information retrieval, fuzzy pattern recognition, image compression and reconstruction, fuzzy automata theory, fuzzy social network analysis and in many other areas.

Sanchez started the study of the linear systems. Solvability and methods for computing the greatest solutions to linear systems of FREIs over various structures of

M. Ćirić (✉) · J. Ignjatović · I. Stanković
Faculty of Sciences and Mathematics, University of Niš, Višegradska 33, 18000 Niš, Serbia
e-mail: miroslav.ciric@pmf.edu.rs

J. Ignjatović
e-mail: jelena.ignjatovic@pmf.edu.rs

I. Stanković
e-mail: ivanstankovic76@gmail.com

© Springer Nature Switzerland AG 2019
M. E. Cornejo et al. (eds.), *Trends in Mathematics and Computational Intelligence*, Studies in Computational Intelligence 796,
https://doi.org/10.1007/978-3-030-00485-9_18

truth values have been investigated in numerous papers. More complex systems of FREIs, called weakly linear, have been recently studied in [13, 15, 16], where the existence of the greatest solutions of these systems has been proved and algorithms for their computing have been provided. Initially, the reason for their study were very important applications in the theory of fuzzy automata, in the state reduction and the study of simulation, bisimulation and equivalence (cf. [6–8, 25]). However, significant applications have also been found in other areas, such as social network analysis, where the solutions of certain weakly linear systems, known as regular equivalences, play a key role in identifying positions of actors in the network.

The problems of social network analysis, namely the problems of positional analysis, have also initiated a study of two-mode system of FREIs, which has been conducted in [26]. In that paper, the existence of the greatest solutions of two-mode systems of FREIs has been proved and algorithms for their computing have been developed. These algorithms simultaneously compute pairs of fuzzy relations which are solutions of the considered two-mode systems.

The main problem which will be considered in this paper also originates from social network analysis. Social network analysis mostly dealt with the one-mode networks, and a wide variety of methods has been developed for handling one-mode networks. However, two-mode and multi-mode networks are also common. Typical examples of two-mode networks include actor-by-event attendance, actor by group membership, actor by trait possession, actor by object possession, and many others. To take advantage of the wealth of methods developed for one-mode networks in the two-mode case, a frequent approach used in work with two-mode networks was their conversion to the one-mode case. Common conversion methods were the method of projections and the method called here the *method of unification*, which comprises treating a two-mode network as a one-mode network over the union of the modes.

The aim of the paper is to study relationships between solutions of two-mode systems of FREIs and solutions of related one-mode systems of FREIs, obtained by the conversion of the two-mode case into the one-mode case by means of the method of unification. We prove that solutions of any two-mode system of FREIs can be derived from solutions of the related one-mode system, and vice versa. This is interesting from the theoretical point of view, but from the computational point of view, the conversion based method leads to an increase in the size of matrices (representing fuzzy relations), so it is more memory and time demanding, what favors the direct method for solving two-mode systems of FREIs.

For fuzzy relations considered in this paper the structure of membership values will be a complete residuated lattice $\mathcal{L} = (L, \wedge, \vee, \otimes, \rightarrow, 0, 1)$ (for a detailed definition see [13, 15, 26]). A *fuzzy relation* between non-empty sets A and B is any fuzzy subset of $A \times B$, i.e., any function $R : A \times B \rightarrow L$. If $A = B$, we say that R is a fuzzy relation on A. The sets of all fuzzy relations between A and B and on A are denoted by $L^{A \times B}$ and $L^{A \times A}$.

A fuzzy relation $\mathbf{0}_{A \times B} \in L^{A \times B}$ defined by $\mathbf{0}_{A \times B}(a, b) = 0$, for each $(a, b) \in A \times B$, is called the *empty relation* between A and B. For $\mathbf{0}_{A \times A}$ we say that it is the empty relation on A.

For $R \in L^{A \times B}$, its *inverse relation* $R^{-1} \in L^{B \times A}$ is defined by $R^{-1}(b, a) = R(a, b)$, for all $(a, b) \in A \times B$, and for $R \in L^{A \times B}$ and $S \in L^{B \times C}$, their *composition* $R \circ S \in L^{A \times C}$ is defined by

$$R \circ S(a, c) = \bigvee_{b \in B} R(a, b) \otimes R(b, c), \qquad \text{for all } (a, c) \in A \times C.$$

Whenever it is defined, the composition is associative, and $(R \circ S)^{-1} = S^{-1} \circ R^{-1}$. For $R, S \in L^{A \times B}$, we write $R \leqslant S$ if $R(a, b) \leqslant S(a, b)$, for every $(a, b) \in A \times B$.

Let $R \in L^{A \times B}$, $S \in L^{B \times C}$ and $T \in L^{A \times C}$, and let fuzzy relations $R \backslash T \in L^{B \times C}$ and $T/S \in L^{A \times B}$ be defined by

$$R \backslash T(b, c) = \bigwedge_{a \in A} R(a, b) \to T(a, c), \qquad T/S(a, b) = \bigwedge_{c \in C} S(b, c) \to T(a, c),$$

for all $(b, c) \in B \times C$ and $(a, b) \in A \times B$. Then $R \backslash T$ is called the *right residual* of T by R, and T/S is called the *left residual* of T by S. It should be noted that $R \backslash T$ is the greatest solution to $R \circ U \leqslant T$, and T/S is the greatest solution to $V \circ S \leqslant T$, where U and V are unknowns which take values in $L^{B \times C}$ and $L^{A \times B}$, respectively. Consequently, $R \circ S \leqslant T \Leftrightarrow S \leqslant R \backslash T \Leftrightarrow R \leqslant T/S$.

For $R \in L^{A \times A}$, if $R(a, a) = 1$, for each $a \in A$, then we say that R is *reflexive*, if $R(a, b) = R(b, a)$, for all $a, b \in A$, then R is *symmetric*, and if $R(a, b) \otimes R(b, c) \leqslant R(a, c)$, for all $a, b, c \in A$, then it is *transitive*. A reflexive and symmetric fuzzy relation is called a *fuzzy quasi-order*, and a reflexive, symmetric and transitive fuzzy relation is called a *fuzzy equivalence*.

For a fuzzy relation $R \in L^{A \times B}$ and $X \subseteq A \times B$, by R_X we denote the restriction of R to X. If the sets A and B are represented as $A = D \cup E$ and $B = F \cup G$, where $D \cap E = F \cap G = \emptyset$, then the expression

$$R = \begin{bmatrix} R_{D \times F} & R_{D \times G} \\ R_{E \times F} & R_{E \times G} \end{bmatrix} \tag{1}$$

is called the *block representation* of R, with blocks $R_{D \times F}$, $R_{D \times G}$, $R_{E \times F}$ and $R_{E \times G}$. If, in addition, $C = I \cup J$, where $I \cap J = \emptyset$, and $S \in \mathbb{L}^{B \times C}$, then we have that

$$R \circ S = \begin{bmatrix} R_{D \times F} & R_{D \times G} \\ R_{E \times F} & R_{E \times G} \end{bmatrix} \circ \begin{bmatrix} S_{F \times I} & S_{F \times J} \\ S_{G \times I} & S_{G \times J} \end{bmatrix} \tag{2}$$

$$= \begin{bmatrix} R_{D \times F} \circ S_{F \times I} \vee R_{D \times G} \circ S_{G \times I} & R_{D \times F} \circ S_{F \times J} \vee R_{D \times G} \circ S_{G \times J} \\ R_{E \times F} \circ S_{F \times I} \vee R_{E \times G} \circ S_{G \times I} & R_{E \times F} \circ S_{F \times J} \vee R_{E \times G} \circ S_{G \times J} \end{bmatrix}$$

Let us consider fuzzy relation equations

(MP3) $(R \circ V)^{-1} = R \circ V$,
(MP4) $(V \circ R)^{-1} = V \circ R$;

where V is an unknown taking values in $L^{B \times A}$ and $R \in L^{A \times B}$ is a given fuzzy relation. Equations (MP3) and (MP4) are two of the four equations which are known as *Moore-Penrose equations*. Moore-Penrose equations and related concepts have been studied in the context of matrices and linear operators, as well as within abstract algebraic structures such as semigroups and rings. They have very important applications in many areas, such as linear algebra, functional analysis, probability, statistics, etc. A fuzzy relation which is the solution of both equations (MP3) and (MP4) is called a $\{3, 4\}$-*inverse* of R. If $\mathcal{R} \subset L^{A \times B}$ is a family of fuzzy relations, a fuzzy relation which is a $\{3, 4\}$-inverse of every $R \in \mathcal{R}$ is called a $\{3, 4\}$-inverse of the family $\mathcal{R}$. According to the results from [14] (see also [5]), for an arbitrary family $\mathcal{R}$ of fuzzy relations there exists the greatest $\{3, 4\}$-inverse of this family. It can be computed using the methodology developed in [5, 14].

2   Fuzzy Relational Systems and Regular Fuzzy Equivalences and Quasi-Orders

A *one-mode fuzzy relational system* is a pair $\mathcal{O} = (A, \mathcal{R})$, where A is a non-empty set and $\mathcal{R} = \{R_i\}_{i \in I}$ is a family of fuzzy relations on A, whereas a *two-mode fuzzy relational system* is a triple $\mathcal{T} = (A, B, \mathcal{R})$, where A and B are non-empty sets, and $\mathcal{R} = \{R_i\}_{i \in I}$ is a family of fuzzy relations between A and B.

Both types of fuzzy relational systems have numerous interpretations and a wide range of practical applications, and here we mainly have in mind their interpretations in the context of *social network analysis*. A one-mode fuzzy relational system can be interpreted as a *one-mode fuzzy social network*, where A is viewed as a set of *actors* or *individuals*, and fuzzy relations from $\mathcal{R}$ represent various relationships between actors. On the other hand, a two-mode fuzzy relational system can be interpreted as a *two-mode fuzzy social network*, where A and B are usually viewed as sets of *actors* and *events*, or something similar, and fuzzy relations from $\mathcal{R}$ determine the participation of actors in events. There is also another interpretation, where A and B are understood as sets of *objects* and *attributes*, and fuzzy relations from $\mathcal{R}$ assign attributes to objects. This interpretation leads to the concept of a *formal context*, which is studied in the framework of the *formal concept analysis*.

One of the central problems of social network analysis is the problem of identifying the *position* or the *role* of an actor in the network, and the main tool used to solve this problem, within the branch of the social network analysis called *positional analysis*, are *regular equivalences*. It should be noted that a kind of quasi-orders, called here right regular, can be even more successful than regular equivalences, as shown recently in [4]. For more information on social network analysis, especially on the positional analysis, we refer to the book [12] and survey articles [2, 3, 9, 18, 24].

In the context of one-mode fuzzy relational systems, regular fuzzy equivalences can be defined through particular systems of fuzzy relation equations. Let $\mathcal{O} = (A, \mathcal{R})$ be a one-mode fuzzy relational system, where $\mathcal{R} = \{R_i\}_{i \in I}$. In a natural way, this fuzzy relational system defines three systems of fuzzy relation equations and inequalities:

$$\alpha \circ R_i = R_i \circ \alpha, \quad i \in I, \tag{3}$$

$$\alpha \circ R_i \leqslant R_i \circ \alpha, \quad i \in I, \tag{4}$$

$$\alpha \circ R_i \geqslant R_i \circ \alpha, \quad i \in I, \tag{5}$$

where α is an unknown taking values in $L^{A \times A}$. Motivated by terminology from social network analysis, solutions to (3) are called *regular fuzzy relations*, solutions to (4) are called *right regular fuzzy relations*, and solutions to (5) are *left regular fuzzy relations*.

These systems of fuzzy relation equations and inequalities have been studied in [15] (see also [13]). It has been proved that all three systems have the greatest solutions, which are fuzzy quasi-orders, and methods for computing these greatest solutions have been provided.

To obtain the greatest solutions that are fuzzy equivalences, systems (3)–(5) have been slightly modified, and the following systems have been considered:

$$\alpha \circ R_i = R_i \circ \alpha, \quad \alpha^{-1} \circ R_i = R_i \circ \alpha^{-1}, \quad i \in I, \tag{6}$$

$$\alpha \circ R_i \leqslant R_i \circ \alpha, \quad \alpha^{-1} \circ R_i \leqslant R_i \circ \alpha^{-1}, \quad i \in I, \tag{7}$$

$$\alpha \circ R_i \geqslant R_i \circ \alpha, \quad \alpha^{-1} \circ R_i \geqslant R_i \circ \alpha^{-1}, \quad i \in I, \tag{8}$$

where α is an unknown taking values in $L^{A \times A}$. Clearly, a fuzzy equivalence is a solution to (6) if and only if it is a solution to (3), and analogous assertions are valid for (7) and (4), and (8) and (5). It has been also proved in [15] that systems (6)–(8) have the greatest solutions, which are fuzzy equivalences, and methods for computing these greatest solutions have been provided. All the systems (3)–(8) will be called here *one-mode systems* of fuzzy relation equations and inequalities.

Two-mode fuzzy relational systems have been studied in [26], where the following systems of fuzzy relation equations and inequalities, assigned to $\mathcal{T} = (A, B, \mathcal{R})$ with $\mathcal{R} = \{R_i\}_{i \in I}$, have been considered:

$$\alpha \circ R_i = R_i \circ \beta, \quad i \in I, \tag{9}$$

$$\alpha \circ R_i \leqslant R_i \circ \beta, \quad i \in I, \tag{10}$$

$$\alpha \circ R_i \geqslant R_i \circ \beta, \quad i \in I, \tag{11}$$

where α and β are unknowns taking values in $L^{A \times A}$ and $L^{B \times B}$, respectively. Clearly, solutions to these systems are pairs of fuzzy relations on A and B and can be ordered coordinatewise. As in the one-mode case, solutions to (9) are called *pairs of regular fuzzy relations*, solutions to (10) are called *pairs of right regular fuzzy relations*, and solutions to (11) are called *pairs of left regular fuzzy relations*. In [26] the existence of the greatest solutions to these systems has been proved, and methods for their computing have been provided. It has been also shown that the greatest solutions are pairs of fuzzy quasi-orders.

In addition, the following systems have been also considered:

$$\alpha \circ R_i = R_i \circ \beta, \quad \alpha^{-1} \circ R_i = R_i \circ \beta^{-1}, \quad i \in I, \tag{12}$$

$$\alpha \circ R_i \leqslant R_i \circ \beta, \quad \alpha^{-1} \circ R_i \leqslant R_i \circ \beta^{-1}, \quad i \in I, \tag{13}$$

$$\alpha \circ R_i \geqslant R_i \circ \beta, \quad \alpha^{-1} \circ R_i \geqslant R_i \circ \beta^{-1}, \quad i \in I, \tag{14}$$

The greatest solutions to these systems, which always exist, are pairs of fuzzy equivalences, and they are computed using similar methods (cf. [26]).

Systems (9)–(14) will be called here *two-mode systems* of fuzzy relation equations and inequalities.

3   One-Mode Conversions

As we have mentioned, one-mode and two-mode fuzzy relational systems have been intensively investigated in the context of social network analysis, as one-mode and two-mode networks. There are three different approaches to the study of two-mode networks. The first approach is a separate analysis of each mode. It is clear that this approach leads to loss of essential information on the two-mode network, so it is used only in combination with other approaches, as a first step in a deeper analysis. The second approach is the conversion of two-mode networks to one-mode networks, and the third one is the true two-mode approach, which includes, among other things, our method of simultaneous computation of pairs of regular fuzzy equivalences and fuzzy quasi-orders. Here we discuss the relationship between the last two approaches, in the context of fuzzy relational systems and regular fuzzy equivalences and fuzzy quasi-orders.

The first method of conversion from two-mode to one-mode case which we meet in the literature is the *method of projections*. If $\mathcal{T} = (A, B, \mathcal{R})$ is a two-mode fuzzy relational system, with $\mathcal{R} = \{R_i\}_{i \in I}$, then one-mode networks $\mathcal{O}' = (A, \mathcal{R}')$ and $\mathcal{O}'' = (B, \mathcal{R}'')$, where

$$\mathcal{R}' = \{R_i \circ R_i^{-1}\}_{i \in I} \quad \text{and} \quad \mathcal{R}'' = \{R_i^{-1} \circ R_i\}_{i \in I},$$

are called *projections* of $\mathcal{T}$. Projections have been widely used in the analysis of two-mode networks (c.f., e.g., [3, 10, 17, 27]), but do not give good results in the positional analysis because there is an example of a pair consisting of the greatest regular fuzzy equivalences on the projections $\mathcal{O}'$ and $\mathcal{O}''$ which is not necessary a pair of regular fuzzy equivalences on the original two-mode network $\mathcal{T}$. Also, it was proved in [17] that the number of links in the projections can be enormously larger than the number of links in the original network, so the projection approach can significantly increase memory and computational complexity.

Here, we are interested in a different approach, where a two-mode fuzzy relational system is treated as a one-mode fuzzy relational system over the union of modes. We call this approach the *method of unification*. Let $\mathcal{T} = (A, B, \mathcal{R})$ be a two-mode fuzzy relational system, where $\mathcal{R} = \{R_i\}_{i \in I}$. We define a one-mode fuzzy relational system $\mathcal{O} = (A \cup B, \mathcal{R}')$ with a family $\mathcal{R}' = \{R_i'\}_{i \in I}$ of fuzzy relations on $A \cup B$ given by the block representations

$$R_i' = \begin{bmatrix} \mathbf{0}_{A \times A} & R_i \\ R_i^{-1} & \mathbf{0}_{B \times B} \end{bmatrix}, \qquad \text{for each } i \in I. \tag{15}$$

Such a way of converting two-mode networks to one-mode networks can be meet, for instance, in [1, 11, 19, 20].

The following theorem explains relationships between regular fuzzy equivalences on $\mathcal{O}$ and pairs of regular fuzzy equivalences on $\mathcal{T}$.

Theorem 1 *Let* $\mathcal{T} = (A, B, \mathcal{R})$ *be a two-mode fuzzy relational system, with* $\mathcal{R} = \{R_i\}_{i \in I}$, *let* $\mathcal{O} = (A \cup B, \mathcal{R}')$ *be the one-mode conversion of* $\mathcal{T}$ *specified by (15), and let* μ *be a fuzzy equivalence on* $A \cup B$ *with the block representation*

$$\mu = \begin{bmatrix} \mu_{A \times A} & \mu_{A \times B} \\ \mu_{B \times A} & \mu_{B \times B} \end{bmatrix}. \tag{16}$$

Then μ *is the greatest regular fuzzy equivalence on* $\mathcal{O}$ *if and only if the following statements are true*

(A) $(\mu_{A \times A}, \mu_{B \times B})$ *is the greatest pair of regular fuzzy equivalences on* $\mathcal{T}$;
(B) $\mu_{B \times A}$ *is the greatest* $\{3, 4\}$*-inverse of the family* $\mathcal{R}$.

Proof Seeing that μ is a fuzzy equivalence, it can be easily shown that $\mu_{A \times A}$ and $\mu_{B \times B}$ are fuzzy equivalences and $\mu_{A \times B} = \mu_{B \times A}^{-1}$.

For each $i \in I$ we have that $\mu \circ R_i' = R_i' \circ \mu$ is equivalent to

$$\begin{bmatrix} \mu_{B \times A}^{-1} \circ R_i^{-1} & \mu_{A \times A} \circ R_i \\ \mu_{B \times B} \circ R_i^{-1} & \mu_{B \times A} \circ R_i \end{bmatrix} = \begin{bmatrix} R_i \circ \mu_{B \times A} & R_i \circ \mu_{B \times B} \\ R_i^{-1} \circ \mu_{A \times A} & R_i^{-1} \circ \mu_{B \times A}^{-1} \end{bmatrix},$$

and this is equivalent to

$$\mu_{A \times A} \circ R_i = R_i \circ \mu_{B \times B}, \tag{17}$$

$$\mu_{A \times A}^{-1} \circ R_i = (R_i^{-1} \circ \mu_{A \times A})^{-1} = (\mu_{B \times B} \circ R_i^{-1})^{-1} = R_i \circ \mu_{B \times B}^{-1}, \tag{18}$$

$$(R_i \circ \mu_{B \times A})^{-1} = \mu_{B \times A}^{-1} \circ R_i^{-1} = R_i \circ \mu_{B \times A}, \tag{19}$$

$$(\mu_{B \times A} \circ R_i)^{-1} = R_i^{-1} \circ \mu_{B \times A}^{-1} = \mu_{B \times A} \circ R_i. \tag{20}$$

Clearly, (17) and (18) mean that $(\mu_{A\times A}, \mu_{B\times B})$ is a pair of regular fuzzy equivalences on $\mathcal{T}$, and (19) and (20) mean that $\mu_{B\times A}$ is a $\{3, 4\}$-inverse of the family $\mathcal{R}$. Therefore, we have proved that μ is a regular fuzzy equivalence on $\mathcal{O}$ if and only if $(\mu_{A\times A}, \mu_{B\times B})$ is a pair of regular fuzzy equivalences on $\mathcal{T}$ and $\mu_{B\times A}$ is a $\{3, 4\}$-inverse of the family $\mathcal{R}$.

Suppose that μ is the greatest regular fuzzy equivalence on $\mathcal{O}$. Let (λ, ϱ) be the greatest pair of regular fuzzy equivalences on $\mathcal{T}$, and let $V \in L^{B\times A}$ be the greatest $\{3, 4\}$-inverse of the family $\mathcal{R}$. It is easy to verify that a fuzzy equivalence on $A \cup B$ with the block representation

$$\begin{bmatrix} \lambda & V^{-1} \\ V & \varrho \end{bmatrix}$$

is a regular fuzzy equivalence on $\mathcal{O}$, whence

$$\begin{bmatrix} \lambda & V^{-1} \\ V & \varrho \end{bmatrix} \leqslant \begin{bmatrix} \mu_{A\times A} & \mu_{B\times A}^{-1} \\ \mu_{B\times A} & \mu_{B\times B} \end{bmatrix},$$

and this implies

$$\lambda \leqslant \mu_{A\times A}, \quad \varrho \leqslant \mu_{B\times B}, \quad V \leqslant \mu_{B\times A}. \tag{21}$$

According to the starting hypotheses on λ, ϱ and V, the inequalities opposite to the inequalities in (21) also hold, so these inequalities are turned into equalities. This means that $(\mu_{A\times A}, \mu_{B\times B})$ is the greatest pair of regular fuzzy equivalences on $\mathcal{T}$ and $\mu_{B\times A}$ is the greatest $\{3, 4\}$-inverse of the family $\mathcal{R}$.

Conversely, let $(\mu_{A\times A}, \mu_{B\times B})$ be the greatest pair of regular fuzzy equivalences on $\mathcal{T}$ and $\mu_{B\times A}$ the greatest $\{3, 4\}$-inverse of the family $\mathcal{R}$. Let θ be an arbitrary regular fuzzy equivalence on $\mathcal{O}$, and assume that

$$\theta = \begin{bmatrix} \theta_{A\times A} & \theta_{B\times A}^{-1} \\ \theta_{B\times A} & \theta_{B\times B} \end{bmatrix}$$

is its block representation. As we have already shown, $(\theta_{A\times A}, \theta_{B\times B})$ is a pair of regular fuzzy equivalences on $\mathcal{T}$ and $\theta_{B\times A}$ is a $\{3, 4\}$-inverse of the family $\mathcal{R}$, which implies

$$\theta_{A\times A} \leqslant \mu_{A\times A}, \quad \theta_{B\times B} \leqslant \mu_{B\times B}, \quad \theta_{B\times A} \leqslant \mu_{B\times A}, \tag{22}$$

so $\theta \leqslant \mu$. Therefore, μ is the greatest regular fuzzy equivalence on $\mathcal{O}$. $\square$

According to the previous theorem, the greatest pair of regular fuzzy equivalences on a two-mode fuzzy relational system $\mathcal{T}$ can be computed through the greatest regular fuzzy equivalence on the related one-mode fuzzy relational system $\mathcal{O}$. Additionally, it is also possible to compute the greatest regular fuzzy equivalence on $\mathcal{O}$ through the greatest pair of regular fuzzy equivalences on $\mathcal{T}$ and the greatest $\{3, 4\}$-inverse of the family $\mathcal{R}$. However, this approach is only interesting from a theoretical point of view, because its computational use leads to an increase in the size of matrices (representing relations), which further increases the memory and computational complexity. From this aspect, the direct method of simultaneous computing the greatest pair of regular fuzzy equivalences on $\mathcal{T}$ is a better solution.

It can be easily verified that the conversion method defined by (15) does not work when we want to link solutions to systems (3) and (12), (4) and (10), (5) and (11), (7) and (13), or (8) and (14). For this purpose we have to slightly modify the definition of the one-mode fuzzy relational system $\mathcal{O} = (A \cup B, \mathcal{R}')$. Namely, instead by (15), we define the family $\mathcal{R}' = \{R_i'\}_{i \in I}$ of fuzzy relations on $A \cup B$ by

$$R_i' = \begin{bmatrix} \mathbf{0}_{A \times A} & R_i \\ \mathbf{0}_{B \times A} & \mathbf{0}_{B \times B} \end{bmatrix}, \quad \text{for each } i \in I. \tag{23}$$

We have that the following theorem is true.

Theorem 2 *Let $\mathcal{T} = (A, B, \mathcal{R})$ be a two-mode fuzzy relational system, with $\mathcal{R} = \{R_i\}_{i \in I}$, let $\mathcal{O} = (A \cup B, \mathcal{R}')$ be the one-mode conversion of $\mathcal{T}$ specified by (23), and let ξ be a fuzzy quasi-order on $A \cup B$ with the block representation*

$$\xi = \begin{bmatrix} \xi_{A \times A} & \xi_{A \times B} \\ \xi_{B \times A} & \xi_{B \times B} \end{bmatrix}. \tag{24}$$

Then ξ is the greatest regular fuzzy quasi-order on $\mathcal{O}$ if and only if the following statements are true

(A) *$(\xi_{A \times A}, \xi_{B \times B})$ is the greatest pair of regular fuzzy quasi-orders on $\mathcal{T}$;*
(B) *$\xi_{B \times A}$ is the greatest solution to the system of fuzzy relation equations*

$$R_i \circ U = \mathbf{0}_{A \times A}, \quad U \circ R_i = \mathbf{0}_{B \times B}, \quad (i \in I), \tag{25}$$

where U is an unknown taking values in $L^{B \times A}$.

Proof This theorem can be proved in a similar way as Theorem 1.

In addition, if ξ is a fuzzy equivalence, then it i clear that ξ is the greatest regular fuzzy equivalence on $\mathcal{O}$ if and only if $(\xi_{A \times A}, \xi_{B \times B})$ is the greatest pair of regular fuzzy equivalences on $\mathcal{T}$ and $\xi_{B \times A}$ is the greatest solution to (25). We can also easily show that the conversion method defined by (23) in the same way links solutions to systems (4) and (10), (5) and (11), (7) and (13), and (8) and (14).

It should also be noted that the greatest solution to (25) is the fuzzy relation from $L^{B \times A}$ represented in terms of residuals as follows:

$$\bigwedge_{i \in I} (R_i \backslash \mathbf{0}_{A \times A}) \wedge (\mathbf{0}_{B \times B} / R_i).$$

References

1. Barber, M.J.: Modularity and community detection in bipartite networks. Phys. Rev. E **76**, 066102 (2007)
2. Batagelj, V.: Social network analysis, large-scale. In: Meyers, R.A. (ed.) Encyclopedia of Complexity and Systems Science, pp. 8245–8265. Springer, New York (2009)
3. Borgatti, S.P.: Two-mode concepts in social network analysis. In: Meyers, R.A. (ed.) Encyclopedia of Complexity and Systems Science, pp. 8279–8291. Springer, New York (2009)
4. Brynielsson, J., Kaati, L., Svenson, P.: Social positions and simulation relations. Soc. Netw. Anal. Min. **2**, 39–52 (2012)
5. Ćirić, M., Ignjatović, J.: The existence of generalized inverses of fuzzy matrices. In: Kóczy, L., Kacprzyk, J., Medina, J. (Eds.), ESCIM (2016). Studies in Computational Intelligence, to appear
6. Ćirić, M., Ignjatović, J., Damljanović, N., Bašić, M.: Bisimulations for fuzzy automata. Fuzzy Sets Syst. **186**, 100–139 (2012)
7. Ćirić, M., Ignjatović, J., Jančić, Damljanović, N.: Computation of the greatest simulations and bisimulations between fuzzy automata. Fuzzy Sets Syst. **208**, 22–42 (2012)
8. Ćirić, M., Stamenković, A., Ignjatović, J., Petković, T.: Fuzzy relation equations and reduction of fuzzy automata. J. Comput. Syst. Sci. **76**, 609–633 (2010)
9. Doreian, P.: Positional analysis and blockmodeling. In: Meyers, R.A. (ed.) Encyclopedia of Complexity and Systems Science, pp. 6913–6927. Springer, New York (2009)
10. Everett, M.G., Borgatti, S.P.: The dual-projection approach for two-mode networks. Soc. Netw. **35**, 204–210 (2013)
11. Fararo, T., Doreian, P.: Tripartite structural analysis: generalizing the Breiger-Wilson formalism. Soc. Netw. **6**, 141–175 (1984)
12. Hanneman, R.A., Riddle, M.: Introduction to Social Network Methods. University of California, Riverside, Riverside CA (2005)
13. Ignjatović, J., Ćirić, M.: Weakly linear systems of fuzzy relation inequalities and their applications: abrief survey. Filomat **26**(2), 207–241 (2012)
14. Ignjatović, J., Ćirić, M.: Moore-Penrose equations in involutive residuated semigroups and involutive quantales. Filomat **31**(2), 183–196 (2017)
15. Ignjatović, J., Ćirić, M., Bogdanović, S.: On the greatest solutions to weakly linear systems of fuzzy relation inequalities and equations. Fuzzy Sets Syst. **161**, 3081–3113 (2010)
16. Ignjatović, J., Ćirić, M., Damljanović, N., Jančić, I.: Weakly linear systems of fuzzy relation inequalities: the heterogeneous case. Fuzzy Sets Syst. **199**, 64–91 (2012)
17. Latapy, M., Magnien, C., Del Vecchio, N.: Basic notions for the analysis of large twomode networks. Soc. Netw. **30**, 31–48 (2008)
18. Lerner, J.: Role assignments. In: Brandes, U., Erlebach, T. (Eds.), Network Analysis: Methodological Foundations, Lecture Notes in Computer Science, vol. 3418, pp. 216–252. Springer (2005)
19. Melamed, D.: Community structures in bipartite networks: a dual-projection approach. PLoS ONE **9**(5), e97823 (2014)
20. Melamed, D., Breiger, R.L., West, A.J.: Community structure in multi-mode networks: Applying an eigenspectrum approach. Connections **33**, 18–23 (2013)
21. Sanchez, E.: Equations de relations Floues. Thèse de Doctorat, Faculté de Médecine de Marseille (1974)
22. Sanchez, E.: Resolution of composite fuzzy relation equations. Inf. Control **30**, 38–48 (1976)
23. Sanchez, E.: Solutions in composite fuzzy relation equations: application to medical diagnosis in Brouwerian logic. In: Gupta, M.M., Saridis, G.N., Gaines, B.R. (Eds.), Fuzzy Automata and Decision Processes, North-Holland, Amsterdam, pp. 221–234 (1977)
24. Scott, J.: Social network analysis, overview of. In: Meyers, R.A. (ed.) Encyclopedia of Complexity and Systems Science, pp. 8265–8279. Springer, New York (2009)
25. Stamenković, A., Ćirić, M., Ignjatović, J.: Reduction of fuzzy automata by means of fuzzy quasi-orders. Inf. Sci. **275**, 168–198 (2014)

26. Stanković, I., Ćirić, M., Ignjatović, J.: Fuzzy relation equations and inequalities with two unknowns and their applications. Fuzzy Sets Syst. **322**, 86–105 (2017)
27. Zhou, T., Ren, J., Medo, M., Zhang, Y.-C.: Bipartite network projection and personal recommendation, Phys. Rev. E 76(4) (2007). Article number 046115

An Adjoint Pair for Intuitionistic *L*-Fuzzy Values

O. Krídlo and M. Ojeda-Aciego

Abstract We continue our prospective study of the generalization of formal concept analysis in terms of intuitionistic *L*-fuzzy sets. The main contribution here is an adjoint pair in the set $\mathcal{L}_{\text{ILF}}$ of intuitionistic *L*-fuzzy values associated to a complete residuated lattice $\mathcal{L}$, which allows the definition of a pair of derivation operators which form an antitone Galois connection.

Keywords Formal concept analysis · Complete residuated lattice
Atanassov's intuitionistic fuzzy sets

1 Introduction

In this work, we continue our study of the extension of Formal Concept Analysis (FCA) to the so-called Atanassov's Intuitionistic fuzzy sets (IF-sets), introduced in [1] by considering for all element x a membership degree $\mu(x)$ together with a non-membership degree $\nu(x)$ such that $\mu(x) + \nu(x) \leq 1$, somehow allowing an *indetermination degree* about x in the case of strict inequality. This construction was later generalized when allowing a complete residuated lattice instead of the unit interval as underlying set of truth-values [2, 5]. Although some authors have already

Partially supported by the Slovak Research and Development Agency contract No. APVV-15-0091, University Science Park TECHNICOM for Innovation Applications Supported by Knowledge Technology, ITMS: 26220220182 and II. phase, ITMS2014+: 313011D232, supported by the ERDF.

Partially supported by the Spanish Science Ministry project TIN15-70266-C2-P-1, co-funded by the European Regional Development Fund (ERDF).

O. Krídlo
University of Pavol Jozef Šafárik, Košice, Slovakia
e-mail: o.kridlo@gmail.com

M. Ojeda-Aciego (✉)
Departamento de Matemática Aplicada, Universidad de Málaga, Málaga, Spain
e-mail: aciego@uma.es

© Springer Nature Switzerland AG 2019
M. E. Cornejo et al. (eds.), *Trends in Mathematics and Computational
Intelligence*, Studies in Computational Intelligence 796,
https://doi.org/10.1007/978-3-030-00485-9_19

introduced intuitionistic extensions of FCA (for instance [10, 12] or [11]), all of
them are based on the unit interval.

In [7], we introduced for the first time a definition of concept-forming operators
purely based on Atanassov's intuitionistic L-fuzzy sets valued on a complete residuated lattice. In order to get a Galois connection in the antitone case, the ILF-formal
context had to provide values without indetermination, i.e. $\mu(x) = \neg(\mu(x))$, which
are essentially equivalent to (usual) L-fuzzy sets. Then in [8] an alternative approach
was presented, in terms of isotone Galois connection and an adjoint triple.

In this paper, we construct an adjoint pair in order to generate (by standard means)
a Galois connection in the set of intuitionistic L-fuzzy sets which, contrariwise to
[7], need not be indetermination-free.

2   Preliminary Definitions

As stated above, we will be primary dealing with truth-values not necessarily belonging to the unit interval, but to a complete residuated lattice (see [6] for further details).

Definition 1 An algebra $\mathcal{L} = \langle L, \wedge, \vee, 0, 1, \otimes, \rightarrow \rangle$ is said to be a *complete residuated lattice* if

1. $\langle L, \wedge, \vee, 0, 1 \rangle$ is a complete lattice where 0 and 1 are the bottom and top elements
   (resp.).
2. $\langle L, \otimes, 1 \rangle$ is a commutative monoid.
3. $\langle \otimes, \rightarrow \rangle$ is an adjoint pair, i.e. $k \otimes m \leq n$ if and only if $k \leq m \rightarrow n$, for all
   $k, m, n \in L$, where $\leq$ is the ordering generated by $\wedge$ and $\vee$.

Let us recall the notion of intuitionistic fuzzy set defined on a complete lattice, as
introduced in [2].

Definition 2 Given a complete lattice L together with an involutive order reversing
operation $N: L \rightarrow L$, and a universe set E: An intuitionistic L-fuzzy set (ILF set)
A in E is defined as an object having the form:

$$A = \{\langle \mu_A(x), \nu_A(x) \rangle / x \mid x \in E\}$$

where the functions $\mu_A: E \rightarrow L$ and $\nu_A: E \rightarrow L$ define the degree of membership
and the degree of non-membership, respectively, to A of the elements $x \in E$, and for
every $x \in E$:

$$\mu_A(x) \leq N(\nu_A(x)).$$

When the previous inequality is strict, there is a certain indetermination degree on
the knowledge about x.

Note that, when the underlying lattice is residuated, we already have a negation operator defined by $\neg x = x \to 0$. As a result, we can define the ILF-lattice associated with a given residuated lattice $\mathcal{L}$ as follows:

Definition 3 Given a complete residuated lattice $\mathcal{L} = \langle L, \wedge, \vee, 0, 1, \otimes, \to \rangle$, we can consider the lattice of intuitionistic truth values

$$\mathcal{L}_{\mathrm{ILF}} = \langle \{\langle k_1, k_2 \rangle \in L \times L \mid k_2 \leq \neg k_1\}, \leq \rangle$$

where ordering $\leq$ on $\mathcal{L}_{\mathrm{ILF}}$ is defined as follows $\langle k_1, k_2 \rangle \leq \langle m_1, m_2 \rangle$ when $k_1 \leq m_1$ and $k_2 \geq m_2$.

Note that $\mathcal{L}_{\mathrm{ILF}}$ is just the construction of the Pareto ordering, as used in [4], considering $\mathcal{L}$ as the underlying set of truth-values instead of the unit interval. Consider also the following notation for any element of $\mathcal{L}_{\mathrm{ILF}}$ as follows $\overline{a} = \langle a_1, a_2 \rangle$.

Lemma 1 $\langle \mathcal{L}_{\mathrm{ILF}}, \leq \rangle$ *forms a complete lattice in which the meet and join are defined by*

$$\bigwedge_{i \in I} \overline{a_i} = \left\langle \bigwedge_{i \in I} a_{i1}; \bigvee_{i \in I} a_{i2} \right\rangle \qquad \bigvee_{i \in I} \overline{a_i} = \left\langle \bigvee_{i \in I} a_{i1}; \bigwedge_{i \in I} a_{i2} \right\rangle$$

Proof It is enough to check that the above defined meet and join actually are elements of $\mathcal{L}_{\mathrm{ILF}}$, since the rest is straightforward.

Given $\{\overline{a_i} \mid i \in I\} \subseteq \mathcal{L}_{\mathrm{ILF}}$, recall that for any $\overline{a_i} \in \mathcal{L}_{\mathrm{ILF}}$ it holds that $a_{i2} \leq \neg a_{i1}$. Hence $\bigwedge_{i \in I} a_{i2} \leq \bigwedge_{i \in I} \neg a_{i1} = \bigwedge_{i \in I} (a_{i1} \to 0) = (\bigvee_{i \in I} a_{i1} \to 0) = \neg \bigvee_{i \in I} a_{i1}$.

On the other hand, we also have that $a_{i1} \leq \neg a_{i2}$ for all $i \in I$. Hence $\bigwedge_{i \in I} a_{i1} \leq \bigwedge_{i \in I} \neg a_{i2} = \neg \bigvee_{i \in I} a_{i2}$, which is equivalent to $\bigvee_{i \in I} a_{i2} \leq \neg \bigwedge_{i \in I} a_{i1}$. $\square$

The definition of the conjunctor in $\mathcal{L}_{\mathrm{ILF}}$ (to be introduced in the next section) will make use of the following operator:

Definition 4 The operator $\oplus \colon L \times L \to L$ is defined by

$$a \oplus b = \neg a \to b = (a \to 0) \to b.$$

Assuming an involutive negation, it is not difficult to check the De Morgan laws between $\otimes$ and $\oplus$, contraposition, and associativity and commutativity of $\oplus$:

Lemma 2 *The following equalities hold*

$$\neg(a \otimes b) = \neg a \oplus \neg b \qquad \neg(a \oplus b) = \neg a \otimes \neg b \qquad a \to b = \neg b \to \neg a$$

Proof We proceed by straightforward calculation as follows:

$$\neg(a \otimes b) = (a \otimes b) \to 0 = a \to (b \to 0)$$
$$= a \to \neg b = \neg a \oplus \neg b$$
$$\neg(a \oplus b) = \neg(\neg\neg a \oplus \neg\neg b) = \neg\neg(\neg a \otimes \neg b)$$
$$= \neg a \otimes \neg b$$
$$\neg b \to \neg a = (b \to 0) \to (a \to 0) = ((b \to 0) \otimes a) \to 0$$
$$= (a \otimes (b \to 0)) \to 0 = a \to ((b \to 0) \to 0)$$
$$= a \to \neg\neg b = a \to b$$

$\square$

If we think of $a \to b$ as $\neg a \oplus b$, then it is easy to see that $\neg(a \to b) = (a \otimes \neg b)$.

Lemma 3 *Let $\mathcal{L}$ be a complete residuated lattice endowed with an involutive negation (i.e. $\neg\neg a = a$). Then $\oplus$ is commutative and associative.*

Proof Firstly,

From $a \to b = \neg b \to \neg a$ we obtain commutativity of $\oplus$

$$a \oplus b = \neg a \to b = \neg b \to a = b \oplus a.$$

Associativity is straightforward

$$(a \oplus b) \oplus c = \neg(a \oplus c) \to c = (\neg a \otimes \neg b) \to c$$
$$= \neg a \to (\neg b \to c) = \neg a \to (b \oplus c) = a \oplus (b \oplus c)$$

$\square$

Hereafter we will assume that $\mathcal{L}$ satisfies the double negation law.

3   The Complete Residuated Lattice $\mathcal{L}_{ILF}$

We will define an intuitionistic conjunctor on $\mathcal{L}_{ILF}$ with the help of the operators $\otimes$ and $\oplus$.

Definition 5 Let $\mathcal{L}_{ILF}$ be the ILF-lattice associated to a residuated lattice $\mathcal{L}$. We define two binary operations on $\mathcal{L}_{ILF}$ by

$$\langle a_1; a_2 \rangle \boxtimes \langle b_1; b_2 \rangle = \langle a_1 \otimes b_1; a_2 \oplus b_2 \rangle$$
$$\langle a_1; a_2 \rangle \rightrightarrows \langle b_1; b_2 \rangle = \langle (a_1 \to b_1) \wedge (\neg a_2 \to \neg b_2); (\neg a_2 \otimes b_2) \rangle$$

for all $\langle a_1, a_2 \rangle, \langle b_1, b_2 \rangle \in \mathcal{L}_{ILF}$.

The following lemma shows that both operations are well defined. Formally,

Lemma 4 $\boxtimes$ *and* $\Rightarrow$ *are internal binary operations in* $\mathcal{L}_{\text{ILF}}$.

Proof We have just to check the condition for belonging to $\mathcal{L}_{\text{ILF}}$, namely, the non-membership degree is less or equal than the negation of the membership degree. In the following chain of equalities we will use the De Morgan laws from Lemma 2.

1. $a_2 \leq \neg a_1$ and $b_2 \leq \neg b_1$. Hence because of the monotonicity of $\oplus$ we have $a_2 \oplus b_2 \leq \neg a_1 \oplus \neg b_1 = \neg(a_1 \otimes b_1)$.
2. $\neg(\neg b_2 \otimes c_2) = b_2 \oplus \neg c_2 = \neg b_2 \to \neg c_2 \geq (\neg b_2 \to \neg c_2) \wedge (b_1 \to c_1)$. Hence $\neg b_2 \otimes c_2 \leq \neg((\neg b_2 \to \neg c_2) \wedge (b_1 \to c_1))$. $\qquad\square$

We can now state and prove the main contribution of this work.

Theorem 1 $\langle \mathcal{L}_{\text{ILF}}, \langle 1, 0 \rangle, \langle 0, 1 \rangle, \boxtimes, \Rightarrow \rangle$ *is complete residuated lattice.*

Proof Firstly, $\mathcal{L}_{\text{ILF}}$ is a complete lattice, by Lemma 1.

$\langle \mathcal{L}_{\text{ILF}}, \boxtimes, \langle 1, 0 \rangle \rangle$ forms a commutative monoid. This is straightforward, by Lemma 3 and the definition of $\boxtimes$.

Finally, let us know prove that $\langle \boxtimes, \Rightarrow \rangle$ is an adjoint pair on $\mathcal{L}_{\text{ILF}}$ which, in our case, means the following:

$$\langle a_1 \otimes b_1, a_2 \oplus b_2 \rangle \leq \langle c_1, c_2 \rangle \iff \langle a_1, a_2 \rangle \leq \langle (b_1 \to c_1) \wedge (\neg b_2 \to \neg c_2), \neg b_2 \otimes c_2 \rangle$$

$\Rightarrow$: Let us assume that $\langle a_1 \otimes b_1, a_2 \oplus b_2 \rangle \leq \langle c_1, c_2 \rangle$.

From the second component we have that $a_2 \oplus b_2 \geq c_2$ but $a_2 \oplus b_2 = \neg b_2 \to a_2 \geq c_2$, and that is equivalent to $a_2 \geq \neg b_2 \otimes c_2$.

From the first component we have $a_1 \otimes b_1 \leq c_1$, which is equivalent to $a_1 \leq b_1 \to c_1$. Moreover, using $\langle a_1, a_2 \rangle \in \mathcal{L}_{\text{ILF}}$ and the previous inequality, we obtain $a_1 \leq \neg a_2 \leq \neg(\neg b_2 \otimes c_2) = \neg\neg b_2 \oplus \neg c_2 = \neg b_2 \to \neg c_2$. Hence, $a_1 \leq (b_1 \to c_1) \wedge (\neg b_2 \to \neg c_2)$.

As a result, we obtain

$$\langle a_1, a_2 \rangle \leq \langle (b_1 \to c_1) \wedge (\neg b_2 \to \neg c_2), \neg b_2 \otimes c_2 \rangle$$

$\Leftarrow$: Conversely, let us assume that $\langle a_1, a_2 \rangle \leq \langle (b_1 \to c_1) \wedge (\neg b_2 \to \neg c_2), \neg b_2 \otimes c_2 \rangle$.

From the first component we obtain $a_1 \leq (b_1 \to c_1) \wedge (\neg b_2 \to \neg c_2) \leq b_1 \to c_1$, which is equivalent to $a_1 \otimes b_1 \leq c_1$.

From the second component we have $a_2 \geq \neg b_2 \otimes c_2$, which is equivalent to $\neg b_2 \to a_2 = a_2 \oplus b_2 \geq c_2$. Hence

$$\langle a_1 \otimes b_1, a_2 \oplus b_2 \rangle \leq \langle c_1, c_2 \rangle.$$

$\square$

4 Antitonic ILF Formal Concept Analysis

Theorem 1 is the key to build a consistent version of formal concept analysis interpreted on $\mathcal{L}_{\mathrm{ILF}}$. To begin with, the notion of ILF-formal context is given as follows:

Definition 6 Let $\mathcal{L}$ be a complete residuated lattice and $\mathcal{L}_{\mathrm{ILF}}$ be its associated lattice of ILF degrees. A triple $\langle B, A, r \rangle$, where $r: B \times A \to \mathcal{L}_{\mathrm{ILF}}$, is said to be an *ILF-formal context*.

The definition of the concept-forming operators associated with an ILF-formal context is introduced in the standard way in terms of $\rightrightarrows$.

Definition 7 Let $\mathcal{L}$ be a complete residuated lattice and let $\mathcal{L}_{\mathrm{ILF}}$ be its associated lattice of ILF values. Given an ILF-formal context $\langle B, A, r \rangle$, we define a pair of mappings $\langle \Uparrow, \Downarrow \rangle$ between the intuitionistic $\mathcal{L}_{\mathrm{ILF}}$-fuzzy powersets $\langle \mathcal{L}_{\mathrm{ILF}}{}^B, \subseteq \rangle$ and $\langle \mathcal{L}_{\mathrm{ILF}}{}^A, \subseteq \rangle$ as follows

(a) $\Uparrow f(a) = \bigwedge_{b \in B} (f(b) \rightrightarrows r(b, a))$, for all $f \in \mathcal{L}_{\mathrm{ILF}}{}^B$
(b) $\Downarrow g(b) = \bigwedge_{a \in A} (g(a) \rightrightarrows r(b, a))$, for all $g \in \mathcal{L}_{\mathrm{ILF}}{}^A$.

The pair of mappings $\langle \Uparrow, \Downarrow \rangle$ are the *concept forming operators* for the IF-formal context $\langle B, A, r \rangle$.

Theorem 2 *Let $\mathcal{L}$ be a complete residuated lattice and $\mathcal{L}_{\mathrm{ILF}}$ its associated lattice of intuitionistic degrees. Let $\langle B, A, r \rangle$ be an IF-formal context. Then $\langle \Uparrow, \Downarrow \rangle$ forms a Galois connection between powersets $\langle \mathcal{L}_{\mathrm{ILF}}^B, \subseteq \rangle$ and $\langle \mathcal{L}_{\mathrm{ILF}}^A, \subseteq \rangle$.*

Proof Follows from Theorem 1 and the standard construction on a complete residuated lattice (see, for instance, [3]). $\square$

The notion of concept in this framework follows the standard approach, and is defined as a fixpoint of the Galois connection from Theorem 2. Similarly, the set of concepts can be ordered by the suitable extension of the subset/superset hierarchy.

5 Conclusions and Future Work

An adjoint pair has been defined on the set of ILF values associated to a complete lattice $\mathcal{L}$ and, as a result, an antitone Galois connection can be induced between the powersets of ILF sets. This result improves a previous attempt in which the Galois connection was only obtained under the assumption that the underlying context is indetermination-free (i.e. $\mu(x) + \nu(x) = 1$ in the standard terminology of IF sets).

As future work, we will study the possible existence of different (families of) adjoint pairs so that the multi-adjoint framework of [9] could also be extended to an ILF setting.

References

1. Atanassov, K.: Intuitionistic fuzzy sets. Fuzzy Sets Syst. **20**, 87–96 (1986)
2. Atanassov, K.: Intuitionistic Fuzzy Sets: Theory and Applications. Physica-Verlag (1999)
3. Bělohlávek, R.: Lattice generated by binary fuzzy relations (extended abstract). In: 4th International Conference on Fuzzy Sets Theory and Applications, p. 11 (1998)
4. Cornelis, C., Kerre, E.: Inclusion measures in intuitionistic fuzzy sets. Lecture Notes Artif. Intell. **2711**, 345–356 (2003)
5. Gerstenkorn, T., Tepavčević, A.: Lattice valued intuitionistic fuzzy sets. Cent. Eur. J. Math. **2**(3), 388–398 (2004)
6. Hájek, P.: Metamathematics of Fuzzy Logic. Kluwer Academic (1998)
7. Krídlo, O., Ojeda-Aciego, M.: Extending formal concept analysis using intuitionistic L-fuzzy sets. In: IEEE International Conference on Fuzzy Systems (FUZZ-IEEE'17) (2017). To appear
8. Krídlo, O., Ojeda-Aciego, M.: Towards intuitionistic L-fuzzy formal t-concepts. In: Joint 17th World Congress of International Fuzzy Systems Association and 9th International Conference on Soft Computing and Intelligent Systems (IFSA-SCIS'17) (2017). To appear
9. Medina, J., Ojeda-Aciego, M., Ruiz-Calviño, J.: Formal concept analysis via multi-adjoint concept lattices. Fuzzy Sets Syst. **160**(2), 130–144 (2009)
10. Pang, J., Zhang, X., Xu, W.: Attribute reduction in intuitionistic fuzzy concept lattices. Abstr. Appl. Anal. (2013), article ID 271398. 12 p.
11. Xu, F., Xing, Z.-Y., Yin, H.-D.: Attribute reductions and concept lattices in interval-valued intuitionistic fuzzy rough set theory: construction and properties. J. Intell. Fuzzy Syst. **30**(2), 1231–1242 (2016)
12. Zhou, L.: Formal concept analysis in intuitionistic fuzzy formal context. In: Seventh International Conference on Fuzzy Systems and Knowledge Discovery (FSKD 2010), pp. 2012–2015 (2010)

On Possibilistic Version of Distance Covariance and Correlation

István Á. Harmati and Robert Fullér

Abstract Distance correlation is a relatively new measure of dependence in probability theory and statistics, which has the great advantage that it gives zero if and only if the variables are independent. In this paper we define its possibilistic version. Namely, we equip each γ-level set of the joint possibility distribution with a uniform probability distribution, then we determine the probabilistic distance covariance and correlation between the marginal distributions. Finally, the possibilistic distance covariance and correlation is computed as the weighted average of these probabilistic measures of dependence.

Keywords Distance correlation · Distance covariance · Possibilistic correlation
Fuzzy numbers · Dependence

1 Introduction

Measuring independence and quantification of the level of dependence plays a fundamental role in probabiliy theory and statistics, but defining a suitable and comfortable measure is not an easy task. The mathematical properties which a 'good' dependence measure should satisfy were listed by Rényi [1], and unfortunatelly the most popular ones, for example Pearson's correlation coefficient, correlation ratio, mutual information do not satisfy all of them. The recently introduced distance correlation [7] is a promising one (sometimes called the correlation coefficient of the 21st century), although its name is quite missleading, since it is not the correlation of distances.

I. Á. Harmati (✉)
Department of Mathematics and Computational Sciences, Széchenyi István University,
Győr, Hungary
e-mail: harmati@sze.hu

R. Fullér
Department of Informatics, Széchenyi István University, Győr, Hungary
e-mail: rfuller@sze.hu

© Springer Nature Switzerland AG 2019
M. E. Cornejo et al. (eds.), *Trends in Mathematics and Computational
Intelligence*, Studies in Computational Intelligence 796,
https://doi.org/10.1007/978-3-030-00485-9_20

The notion of independence has also a crucial role in possibility theory, similarly to probability theory. There are several kinds of independence concepts (see for example [2, 3]) and lot of measures of indepence, see for example [4, 5]. In this article we follow the method introduced by Fullér et al. [13], namely the possibilistic measure of dependence is computed as a weighted average of probabilistic ones defined on the crisp γ-levels sets.

2 Distance Correlation

Distance correlation was introduced by Székely et al. [6, 7], they also built it into the framework of statistical procedures [8, 9].

The number of applications grows fastly, for example in astrophysics [10], time-series analysis [11], neuro science [12].

The motivation is to point out if there are any dependencies between random variables X and Y. It is well-known that they are independent if and only if $f_{XY} = f_X \cdot f_Y$, where f_{XY}, f_X and f_Y are probability density functions (pdf) of (X, Y), X and Y, respectively. This leads to measure the distance between f_{XY} and $f_X \cdot f_Y$, for example by a suitable norm $\| f_{XY} - f_X \cdot f_Y \|$, but the pdfs do not behave nicely, i.e. not always exist and may not be always continuous. So instead of the pdf, it is more promising to use the charatheristic function of the random variables, which has the following form for a p dimensional random vector $\mathbf{X}$:

$$\varphi_{\mathbf{X}}(\mathbf{t}) = E\left(e^{i\langle \mathbf{tx}\rangle}\right) = \int_{\mathbb{R}^p} e^{i\langle \mathbf{tx}\rangle} f_{\mathbf{X}}(\mathbf{x}) \, d\mathbf{x} \tag{1}$$

The most important property is that the characteristic function shares the property of the probability density function, i.e. $\varphi_{\mathbf{XY}} = \varphi_{\mathbf{X}}\varphi_{\mathbf{Y}}$ if and only if random vectors $\mathbf{X} \in \mathbb{R}^p$ and $\mathbf{Y} \in \mathbb{R}^q$ are independent. Based on this fact, the distance covariance is defined as follows:

$$\mathrm{dCov}^2(X, Y) = \int_{\mathbb{R}^{p+q}} |\varphi_{\mathbf{XY}} - \varphi_{\mathbf{X}}\varphi_{\mathbf{Y}}|^2 \, w(t, s) \, dt \, ds \tag{2}$$

Here the rule of weighting function w is to produce a scale and rotation invariant measure that does not equal to zero for dependent random variables. Different choices of w provide different types of covariance measures. In [7] Székely and Rizzo applied the following non-integrable function

$$w(t, s) = \frac{1}{c_p c_q |s|_p^{1+p} |t|_q^{1+q}} \tag{3}$$

where $|\cdot|$ denotes the Euclidean norm, and

$$c_d = \frac{\pi^{\frac{d+1}{2}}}{\Gamma\left(\frac{d+1}{2}\right)} \tag{4}$$

The distance covariance with this weight function is the square root of the following:

$$dCov^2(X, Y) = \frac{1}{c_p c_q} \int_{\mathbb{R}^{p+q}} \frac{\left|\varphi_{X,Y}(s, t) - \varphi_X(s)\varphi_Y(t)\right|^2}{|s|_p^{1+p} |t|_q^{1+q}} \, dt \, ds \tag{5}$$

As in the case of the 'original' covariance, the correlation coefficient is defined by covariances, using that $Var(X) = Cov(X, X)$.

$$dCor(X, Y) = \frac{dCov(X, Y)}{\sqrt{dVar(X)dVar(Y)}} \tag{6}$$

Most important properties of the distance correlation:

1. $0 \le dCor(X, Y) \le 1$ (note that Pearson's correlation can be negative)
2. $dCor(X, Y) = 0$ if and only if X and Y are independent (note that Pearson's correlation can be zero for dependent variables, too).

2.1 Sample Distance Correlation

The previous choice of parameters c_p and c_q (see Eq. 4) has a great advantage that, in this case, we get a nice estimator for the distance covariance (and for the variance also, and finally, for the distance correlation). Let (X_k, Y_k), $k = 1, 2, \ldots, n$ be a sample of random variables (X, Y) (they can be random vectors also). Firstly, we compute the distance matrices a and b, using the Euclidean distance $\|X_j - X_k\|$:

$$a_{j,k} = \|X_j - X_k\|, \qquad \text{for all } j, k = 1, 2, \ldots, n \tag{7}$$
$$b_{j,k} = \|Y_j - Y_k\|, \qquad \text{for all } j, k = 1, 2, \ldots, n \tag{8}$$

Then compute the means of the rows $(\overline{a}_{j.})$, columns $(\overline{a}_{.k})$, and all elements of the distance matrices $(\overline{a}_{..})$. Do the same for b. Define matrices A and B:

$$A_{j,k} = a_{j,k} - \overline{a}_{j*} - \overline{a}_{*k} + \overline{a}_{..} \tag{9}$$
$$B_{j,k} = b_{j,k} - \overline{b}_{j*} - \overline{b}_{*k} + \overline{b}_{..} \tag{10}$$

Finally, the sample distance covariance can be computed from the following expression:

$$dCov_n^2(X, Y) = \frac{1}{n^2} \sum_{j,k=1}^{n} A_{j,k} B_{j,k} \tag{11}$$

The sample distance correlation (using that $\mathrm{Var}(X) = \mathrm{Cov}(X, X)$):

$$\mathrm{dCor}_n(X, Y) = \frac{\mathrm{dCov}_n(X, Y)}{\sqrt{\mathrm{dVar}_n(X)\mathrm{dVar}_n(Y)}} \tag{12}$$

3 Possibilistic Distance Correlation

A fuzzy number A is a fuzzy set of $\mathbb{R}$ with a normal, fuzzy convex and continuous membership function of bounded support. Fuzzy numbers can be considered as possibility distributions. A fuzzy set C in $\mathbb{R}^2$ is said to be a joint possibility distribution of fuzzy numbers A, B, if it satisfies the relationships

$$\max\{x \mid C(x, y)\} = B(y), \quad and \quad \max\{y \mid C(x, y)\} = A(x), \tag{13}$$

for all $x, y \in \mathbb{R}$. Furthermore, A and B are called the marginal possibility distributions of C. A γ-level set (or γ-cut) of a possibility distribution C is a non-fuzzy set denoted by $[C]^\gamma$ and defined by

$$[C]^\gamma = \begin{cases} \{(x, y) \in \mathbb{R}^2 \mid C(x, y) \geq \gamma\} & \text{if } \gamma > 0 \\ \mathrm{cl}(\mathrm{supp}C) & \text{if } \gamma = 0 \end{cases} \tag{14}$$

where $\mathrm{cl}(\mathrm{supp}C)$ denotes the closure of the support of C.

Fullér, Mezei and Várlaki introduced the definition of possibilistic correlation coefficient [13] between marginal distributions of the joint possibility distribution, as an improvement of the earlier definition in [14]:

Definition 1 Let $f : [0, 1] \to \mathbb{R}$ be a non-negative, monotone increasing function with the normalization property $\int_0^1 f(\gamma)d\gamma = 1$. The f-weighted possibilistic correlation coefficient of fuzzy numbers A and B (with respect to their joint distribution C) is defined by

$$\rho_f(A, B) = \int_0^1 \rho(X_\gamma, Y_\gamma) f(\gamma)d\gamma, \tag{15}$$

where

$$\rho(X_\gamma, Y_\gamma) = \frac{\mathrm{cov}(X_\gamma, Y_\gamma)}{\sqrt{\mathrm{var}(X_\gamma)}\sqrt{\mathrm{var}(Y_\gamma)}},$$

and, where X_γ and Y_γ are random variables whose joint distribution is uniform on $[C]^\gamma$ for all $\gamma \in [0, 1]$, and $\mathrm{cov}(X_\gamma, Y_\gamma)$ denotes their probabilistic covariance.

Applying a similar approach, Fullér et al. defined other possibilistic measures of dependence and determined their values for several type of joint possibility distributions [15, 16]. In this way, we can give the definition of the possibilistic distance correlation (and the possibilistic distance covariance, of course):

Definition 2 Let $f: [0, 1] \to \mathbb{R}$ be a non-negative, monotone increasing function with the normalization property $\int_0^1 f(\gamma)d\gamma = 1$. The f-weighted possibilistic distance correlation coefficient of fuzzy numbers A and B (with respect to their joint distribution C) is defined by

$$\mathrm{dCor}_f(A, B) = \int_0^1 \mathrm{dCor}(X_\gamma, Y_\gamma) f(\gamma) d\gamma, \tag{16}$$

where X_γ and Y_γ are random variables whose joint distribution is uniform on $[C]^\gamma$, for all $\gamma \in [0, 1]$.

4 Examples

In this section, we show examples for the possibilistic distance covariance and correlation, when the joint possibility distribution has a special structure, namely it is defined by a t-norm.

One of the most widely used t-norm is the minimum (min) operator. If the joint possibility distribution is defined by the minimum of the marginal distributions (fuzzy numbers) then it is referred as the case of non-interactivity [3]. Non-interactivity implies zero possibilistic correlation coefficent [13], and it is true for possibilistic distance correlation, too, since in this case any γ-level set can be transformed into the $[0, 1] \times [0, 1]$ domain by shifting and scaling (under these transformations the correlation coefficient is invariant). Using the characteristic functions of the joint and marginal probability distributions, the distance covariance can be determined as follows:

$$\varphi_{X,Y}(s, t) = \int_{-\infty}^{\infty} \int_{-\infty}^{\infty} e^{i(xt+ys)} f(x, y)\, \mathrm{d}x\, \mathrm{d}y = -\frac{(1-e^{it})(1-e^{is})}{st} \tag{17}$$

$$\varphi_X(t) = \int_{-\infty}^{\infty} e^{ixt} f(x)\, \mathrm{d}x = \frac{i \cdot (1-e^{it})}{t} \tag{18}$$

$$\varphi_Y(t) = \int_{-\infty}^{\infty} e^{ist} f(y)\, \mathrm{d}y = \frac{i \cdot (1-e^{is})}{s} \tag{19}$$

It follows that $\left|\varphi_{X,Y}(s, t) - \varphi_X(t) \cdot \varphi_Y(t)\right| = 0$, which implies that dCov $= 0$ and dCor $= 0$ for every level set, so $\mathrm{dCor}_f(A, B) = 0$ for any weighting function.

Our other example is the case when the joint possibility distribution is defined by the Łukasiewicz t-norm, i.e.

$$C(x, y) = \begin{cases} x + y - 1 & \text{if } 0 \leq x, y \leq 1 \text{ and } x + y \geq 0, \\ 0 & \text{otherwise.} \end{cases} \qquad (20)$$

The γ-level sets of C are $[C]^\gamma = \{(x, y) \in \mathbb{R}^2 | \gamma \leq x, y \leq 1, \ x + y \geq 1 + \gamma \}$, which are similar triangles, so it is enough to determine the correlation for the case when $\gamma = 0$. Although this case is yet tractable, we solve it numerically, because for more difficult joint distributions the exact computation of the characteristic function based integral can be really tedious.

We generated random samples from a two dimensional uniform distribution defined on $[C]^0$. The number of samples was 100, the sample size was 1000 in each cases. The mean of the sample distance correlation was 0.4732 (standard deviation 0.0214). This suggest that there is no direct functional relationship between the marginal possibility distributions, but they are not independent.

5 Summary

We introduced the notion of possibilistic distance correlation, applying similar methodology we already used in the generalization of other probabilistic measures of dependence. The exact computation of this value often leads to really difficult formulae, which highly depend on the boundaries of the γ-level sets. Since in applications these boundaries are usually not certain and the very precise value of the correlation is not required, we recommend using the sample distance covariance and correlation, after generating uniformly distributed values on the γ-level sets.

Although the possibilistic distance correlation doesn't have any practical applications yet, as a future work we plan to apply it in analysis of hydrological processes and time series, where stochastic and possibilistic uncertainty also arise.

Acknowledgements This work was supported by EFOP-3.6.2-16-2017-00015, HU-MATHS-IN – Intensification of the activity of the Hungarian Industrial Innovation Service Network.

References

1. Rényi, A.: On measures of dependence. Acta Math. Hungarica **10**(3), 441–451 (1959)
2. de Campos, L.M., Huete, J.F.: Independence concepts in possibility theory: Part I. Fuzzy Sets Syst. **103**, 127–152 (1999). Part II. Fuzzy Sets Syst. **103**, 487–505 (1999)
3. Fullér, R., Majlender, P.: On interactive fuzzy numbers. Fuzzy Sets Syst. **143**, 355–369 (2004). https://doi.org/10.1016/S0165-0114(03)00180-5
4. Liu, S.T., Kao, C.: Fuzzy measures for correlation coefficient of fuzzy numbers. Fuzzy Sets Syst. **128**, 267–275 (2002)

5. Hong, D.H.: Fuzzy measures for a correlation coefficient of fuzzy numbers under T_W (the weakest t-norm)-based fuzzy arithmetic operations. Inf. Sci. **176**, 150–160 (2006)
6. Székely, G.J., Rizzo, M.L., Bakirov, N.K.: Measuring and testing dependence by correlation of distances. Ann. Stat. **35**(6), 2769–2794 (2007)
7. Székely, G.J., Rizzo, M.L.: Brownian distance covariance. Ann. Appl. Stat. **3**(4), 1236–1265 (2009)
8. Székely, G.J., Rizzo, M.L.: The distance correlation t-test of independence in high dimension. J. Multivar. Anal. **117**, 193–213 (2013)
9. Székely, G.J., Rizzo, M.L.: Partial distance correlation with methods for dissimilarities. Ann. Appl. Stat. **42**(6), 2382–2412 (2014)
10. Martínez-Gómez, E., Richards, M.T., St, D., Richards, P.: Distance correlation methods for discovering associations in large astrophysical databases. Astrophys. J. **781**(1), 39 (2014)
11. Zhou, Z.: Measuring nonlinear dependence in time-series, a distance correlation approach. J. Time Ser. Anal. **33**(3), 438–457 (2012)
12. Geerligs, L., Henson, R.N.: Functional connectivity and structural covariance between regions of interest can be measured more accurately using multivariate distance correlation. NeuroImage **135**, 16–31 (2016)
13. Fullér, R., Mezei, J., Várlaki, P.: An improved index of interactivity for fuzzy numbers. Fuzzy Sets Syst. **165**, 50–60 (2011). https://doi.org/10.1016/j.fss.2010.06.001
14. Carlsson, C., Fullér, R., Majlender, P.: On possibilistic correlation. Fuzzy Sets Syst. **155**, 425–445 (2005). https://doi.org/10.1016/j.fss.2005.04.014
15. Fullér, R., Harmati, I.Á., Várlaki, P., Rudas, I.: On weighted possibilistic informational coefficient of correlation. Int. J. Math. Models Methods Appl. Sci. **6**(4), 592–599 (2012)
16. Fullér, R., Harmati, I.Á., Mezei, J., Várlaki, P.: On possibilistic correlation coefficient and ratio for fuzzy numbers. In: Recent Researches in Artificial Intelligence, Knowledge Engineering & Data Bases, 10th WSEAS International Conference on Artificial Intelligence, Knowledge Engineering and Data Bases, February 20–22. Cambridge, UK, WSEAS Press, pp. 263–268 (2011). [ISBN 978-960-474-237-8]

A Fuzzy Approach for Measuring Sentence Checkability—Preliminary Results

Hugo Farinha and Joao P. Carvalho

Abstract Fact-checking has recently become a real world hot topic, especially in what concerns political claims. Several big players, such as, for example, Google or Facebook, have started addressing/making contributions to make "Fact-checking" possible/available to the general public. However, most, if not all Fact-checking platforms are largely manual, in the sense that most of the contributions and of the actual checking is performed by humans. Automatic computational Fact-checking is still very far from being reliable and available on a large scale. In this paper we contribute to the goal of automatic Fact-checking by presenting a fuzzy approach to computing sentence checkability, i.e., to answer the question: "is it possible to know if a sentence is worth to be checked?"

Keywords Automatic Fact-checking · Sentence checkability · Fuzzy sets
Natural language processing

1 Introduction: Is a Sentence Checkable? How Checkable Is It?

With the widespread of the internet, and the social web playing such an important role in peoples' lives, the necessity for automatic ways to validate the veracity of the information circulating in the web is increasing rapidly, and is currently a very hot topic. Automated Fact-checking is the name given to the process of, given a claim, a fact or, in general, a sentence, automatically assess its veracity based on publicly available sources of information in the internet. This paper focuses on one of the

H. Farinha · J. P. Carvalho (✉)
INESC-ID, Lisbon, Portugal
e-mail: joao.carvalho@inesc-id.pt

H. Farinha
e-mail: hugo_miguel96@hotmail.com

H. Farinha · J. P. Carvalho
Instituto Superior Técnico, Universidade de Lisboa, Lisbon, Portugal

© Springer Nature Switzerland AG 2019
M. E. Cornejo et al. (eds.), *Trends in Mathematics and Computational Intelligence*, Studies in Computational Intelligence 796,
https://doi.org/10.1007/978-3-030-00485-9_21

first steps deemed necessary to achieve "Automated Fact-checking": measure the potential for a sentence "checkability".

As an example of a sentence that can be checked, or, its veracity can be assessed, is "Albert Einstein was born in 1879 in Germany". On the opposite side, "What was accomplished yesterday was nearly perfect" can't be checked, because, not only it's not easy to define what is "perfect" and what a "perfect event" is, but also it's impossible to define what a "nearly perfect event" is.

The exact meaning of "what is checkable" or "check-worthy" is not trivial. In this work we opted to consider that "checkable sentences" must be factoids [1] or exhibit similar characteristics. A factoid is a fact that has appeared in the news (in our case, a fact that can be checked in the internet). In this paper we propose an approach based on Natural Language Processing (NLP) and Fuzzy sets to address a way to indicate how checkable a factoid is. The approach measures the potential of the quality of a future fact search and the accuracy of the veracity classification, i.e., the higher the checkability, the higher should be the chance to get a correct classification. As an example, "Albert Einstein was born in 1879" should score less than "Albert Einstein was born in 1879 in Germany", since the search related to the latter will have more relevant information and the classification has more relevant information to check.

2 Related Work

Even though Fact-checking is currently a hot-topic, current technologies for full automated Fact-checking are still either too dependent on "manual labour" or limited in scope.

Most of the research work in Fact-checking addresses search and veracity classification. In this sub-topic it is possible to find several major players using different algorithms and (mostly manual) implementations: FightHoax (based on Artificial Intelligence) [2], WikiTribune (the facts are verified by professional journalists) [3], Google's Fact Check (labels news or claims that are fact-checked by news publishers or Fact-checking organizations) [4] or politifacts (claims classified by editors and reporters from the Tampa Bay Times) [5].

Regarding the identification of check-worthy sentences, works are much sparser, and the only solution with some traction is ClaimBuster, which classifies a sentence's checkability in respect to "how factual it is" and "how relevant to the public it is if checked" [6]. However, ClaimBuster is very limited and its results highly questionable. For example, the claim "Donald Trump plans to build a wall dividing USA and Mexico" scores 22% in ClaimBuster [7]. The "check-worthiness" of a claim is relative and different for each person, but a score of 22% (almost not checkable) seems too low for a well-written and relevant present day claim.

3 Computing a Sentence Checkability

As indicated above, we assume that the sentence to be checked is a factoid, and, a priori, checkable. We propose to compute the sentence checkability based on the sentence's subjectivity, sentiment and opinion level, entropy and use of keywords. We make use of (mostly NLP) tools that are publicly available to compute how the sentence scores regarding those factors, and use expert fuzzy knowledge to aggregate the information. Finally we perform some syntax and semantic analysis to adjust the proposed checkability score. The following sections describe each step of the process.

3.1 Subjectivity and Sentiment/Opinion Score

By analyzing sentences used in claims extracted from articles and news, it was clear that sentiment and the expressing of opinions should be taken into account when computing how checkable a sentence is. For example "Hotel X was very comfortable", reveals a personal opinion and, so, is considerably less checkable than, for say, "Hotel X has 120 rooms". Sentiment Vader [8], from the package NLTK, was used to assess the level of sentiment or opinion on a sentence. Since this method is dedicated to "Opinion Mining", which is not exactly the task we are addressing, we introduced a subjectivity classification in order to adapt and improve the results. NLTK Sentiment Utils [9] was used to perform the subjectivity classification.

The Sentiment/Opinion Score is used only when the sentence is classified as "Subjective" or this score is above a given threshold. Through testing and fine-tuning it was decided to use 0.6 as the Sentiment/Opinion threshold score.

3.2 Entropy and Keywords Score

Two other metrics deemed important to compute the checkability of a sentence are the amount of information within a sentence, and the use and importance of keywords. The more information on the sentence, the more information there is to check and the more definite the entities on the sentence are.

Modified Shannon Entropy. Computing the entropy is the most reasonable step to measure the amount of information on a sentence. Entropy uses probabilities of each symbol, or words in this case, to compute the desired measure. We used Shannon Entropy [10]:

$$P(x_i) = \frac{N_i}{\sum_{k=1}^{n} N_k}, \; with \; N_i = Number \; of \; occurrences \; of \; the \; word_i \quad (1)$$

To improve results, not all of the words on the sentence were used. From the usual Bag of Words we excluded:

- Stop words from NLTK corpus ("the", for example) [11];
- Words (not numbers) with less than 3 characters ("Mr", for example);
- All the punctuation (",", ";","!", etc.).

Keywords Score—Modified RAKE. Finally we quantify the sentence's keywords and their relevance. This score is a "dual" of the Entropy Score since it also measures the information on the sentence. However, with this score we try to measure the relevance of the information instead of the amount of information.

To compute this score, a very simplistic approach was used: the RAKE method [12]. This method identifies keywords by going through the sentence and splitting it to remove stopwords. The method finds keywords sets and computes a score for each set. The final Keywords Score is simply the sum of the sets' scores. Each score is computed based on: the number of words (consecutive non-stopwords) and the number of occurrences of each one on the sentence. To improve results the method was modified as follows:

- Punctuation was removed from the sentence before applying the method;
- Since RAKE does not give, for itself, score to numbers and numbers take a huge part in Fact-checking (dates, values, percentages, amounts, etc.), the RAKE output was modified to 1.5 if the set is a number (i.e. numbers are worth slightly more than one word).

Normalization. Shannon's entropy and RAKE scores can take values between 0 and $+\infty$, and there is no perception of what is an optimum or even a good value in what concerns measuring a sentence checkability. With this in mind, we devised a normalization procedure for both scores using several training sentences containing different amounts of information (manually classified with a desired score). Figure 1 shows the normalization result for the training {computed value, desired score} pairs. The computed entropy was represented in $2^{H(x)}$, with H(x) being the entropy computed using (1). Linear interpolation is applied to obtain a score to any computed value. As a result of the normalization procedure we obtain fuzzifiable scores.

3.3 Fuzzy Score Aggregation

The aggregation of the computed scores is done using expert knowledge expressed by a Fuzzy If-Then Mandani rule base (implemented using [13]). Fuzzy Sets were chosen because of the following reasons:

- Scores are normalized and can take any value between 0 and 1;
- The several thresholds, parameters and functions used to manipulate, adapt and compute the individual scores, were used based on testing and expert knowledge,

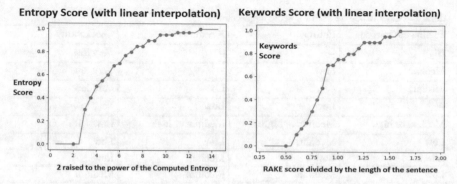

Fig. 1 Normalization of the computed Entropy and RAKE scores

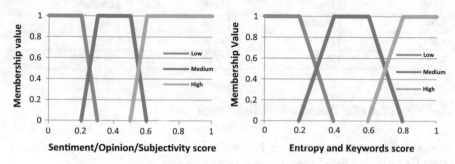

Fig. 2 Membership functions for the three input scores

and are not proven to be optimal. Therefore the method to be used should be robust and not change drastically with further changes of these defined parameters;

- Expert knowledge was easy to implement and gave sound preliminary results.

We defined three membership functions for the inputs (Low, Medium and High). Experts found it difficult to use more than three linguistic terms when devising the rules for this problem. We used five membership functions evenly distributed through the Universe of Discourse (VL, L, M, H, VH) to indicate Checkability. Trapezoidal functions provided the best trade-off between complexity and accuracy of the result (Triangular was found to be slightly worse). The membership functions were defined as represented in Fig. 2.

Table 1 shows the fuzzy rulebase used to aggregate the computed scores. Each line contains one rule of the type "IF Sentiment/Opinion score is A, and Entropy score is B, and Keywords score is C, THEN Checkability is Z", where A, B, C and Z are fuzzy linguistic membership functions. The rules were defined based on expert knowledge and annotated data examples.

Table 1 Fuzzy Rulebase used for score aggregation

Sentiment/Opinion	Entropy	Keywords	Checkability
High	–	–	Very low
Medium	Low	–	Very low
Medium	–	Low	Very low
–	Low	Low	Very low
Medium or high	Medium or high	Medium or high	Low
Low	Medium	Low	Low
Low	Low	Medium	Low
Low	Medium	Medium	Medium
Low	Low	High	Medium
Low	High	Low	Medium
Low	Medium	High	High
Low	High	Medium	High
Low	High	High	Very high

3.4 Syntax and Semantic Adjustment

Some additional metrics based on the syntax and the semantics of the sentence were added in order to further improve the obtained score.

We identify valid/invalid temporal expressions or locations using Semantic Role Labeling [14] based on SENNA [15]. The reasoning behind this procedure is that sentences that refer to a non-existing location or time, are obviously less checkable. After SENNA identifies such expressions, we try to validate them with Ternip (temporal expressions) [16] or a custom database built based on GeoNames [17] and Google Places (locations) [18]. The sentence checkability is penalized by 40% if all expressions are invalid, or increased by 10% if an expression is valid.

Then we check whether after applying Coreference Resolution, performed using CoreNLP [19], the sentence still contains personal pronouns. If affirmative, i.e., CoreNLP could not find the entity to link the pronoun, then the sentence's checkability is decreased by 40%.

Finally we identify indefinite entities, such as, for example "The ship ...". CoreNLP does a syntactic analysis [20] on the sentence that is used to group Noun Phrases (NP). In each group, each common noun is verified for the existence of articles (such as "a") or adjectives (such as "several") before the noun and some words after the noun that can describe it. If there is an article or adjective before the noun but no following words (on the same NP group), the sentence's checkability is decreased by 40%.

4 Preliminary Results, Conclusions and Future Work

The described approach has been implemented and tested in a large range of sentences. Preliminary results are sound, very encouraging and compare very favorably against ClaimBuster. Table 2 shows some examples of the computed checkability.

Proper validation, including ClaimBuster comparisons, using an extensive dataset and external expert evaluators is obviously needed to support the conclusions, and is currently under way.

Despite the interesting results, the proposed approach still has room for improvement, especially in what concerns the optimization of the fuzzy sets and rulebase (which could be automatically tuned once enough data is available after the ongoing validation). Other possible improvements include the fuzzification of the Syntactic and Semantic adjustments.

Table 2 Preliminary results and comparison with ClaimBuster [6]

Sentence	ClaimBuster score (%)	Computed checkability (%)
HSBC lays off 120 technology staff in Hong Kong in cost-cutting plan	37	100
Thomas Edison invented the first commercial light bulb in 1879	35	100
Donald Trump plans to build a wall dividing USA and Mexico	22	91.67
In state after state polls make clear that the American public understands the Kelo ruling is a disaster	36	54
Lisbon has a weak system of public transports	18	27
Seven US Navy crew members are missing after their ship collided with a merchant vessel off the coast of Japan	45	55
Seven US Navy crew members are missing after their ship collided with a merchant vessel off the coast of Japan (Same sentence as above, excluding the period. The score should be equal)	39	55
Donald Trump: It's – the premiums are going up 60, 70, 80%	76	34.78

Acknowledgements Work supported by national funds through Fundação para a Ciência e a Tecnologia (FCT) under reference UID/CEC/50021/2013 and SFRH/BSAB/136312/2018.

References

1. Hiskey, D. (2010). http://www.todayifoundout.com/index.php/2010/02/the-difference-between-a-fact-and-a-factoid/. Accessed 9 June 2017
2. FightHoax Homepage. http://fighthoax.com. Accessed 11 June 2017
3. WikiTribune Homepage. https://www.wikitribune.com. Accessed 11 June 2017
4. Kosslyn, J., Yu, C. (2017) https://blog.google/products/search/fact-check-now-available-google-search-and-news-around-world/. Accessed 11 June 2017
5. Politifact Homepage. http://www.politifact.com/truth-o-meter/. Accessed 11 June 2017
6. Hassan, N. et al.: The quest to automate fact-checking. In: Proceedings of the 2015 Computation+Journalism Symposium (2015)
7. ClaimBuster Demo. http://idir-server2.uta.edu/claimbuster/demo. Accessed 11 June 2017
8. Hutto, C.J., Gilbert, E.E.: VADER: a parsimonious rule-based model for sentiment analysis of social media text. In: Eighth International Conference on Weblogs and Social Media (ICWSM-14). Ann Arbor, MI, June 2014
9. Pang, B., Lee, L.: A sentimental education: sentiment analysis using subjectivity summarization based on minimum cuts. In: Proceedings of the ACL (2004)
10. Wiktionary. https://en.wiktionary.org/wiki/Shannon_entropy. Accessed 20 May 2017
11. NLTK Corpus. http://www.nltk.org/howto/corpus.html. Accessed 16 May 2017
12. Rose, S., Engel, et al.: Automatic keyword extraction from individual documents. In: Berry, M.W., Kogan, J. (eds.) Text Mining: Theory and Applications (2010)
13. Warner, J.D. et al.: JDWarner/scikit-fuzzy: Scikit-Fuzzy 0.3. Zenodo (2017). http://doi.org/10.5281/zenodo.802397
14. Jurafsky, D., Martin, J.H.: Speech and Language Recognition, 3rd edn. draft, Chapter 22 - Semantic Role Labeling (2015)
15. Collobert, R., et al.: Natural language processing (almost) from scratch. J. Mach. Learn. Res. (JMLR) (2011)
16. Ternip—Temporal Expression Recognition and Normalization. https://github.com/cnorthwood/ternip. Accessed 25 May 2017
17. GeoNames Database. http://www.geonames.org/export/. Accessed 25 May 2017
18. GooglePlaces. https://developers.google.com/places/web-service. Accessed 25 May 2017
19. Clark, K., Manning, C.D.: Entity-centric coreference resolution with model stacking. In: Association for Computational Linguistics (2015)
20. Jurafsky, D., Martin, J.H.: Speech and Language Recognition, 3rd edition draft, Chapter 12 – Syntactic Parsing (2015)

Reduced Model Investigations Supported by Fuzzy Cognitive Map to Foster Circular Economy

Adrienn Buruzs, Miklós F. Hatwágner and László T. Kóczy

Abstract The aim of the present paper is to develop an integrated method that provide assistance to decision makers during system planning, design, operation and evaluation. In order to support the realization of Circular Economy (CE) it is essential to evaluate local needs and conditions that help to select the most appropriate system-components and resources needed. Each of these activities requires careful planning, however, the model of CE offers a comprehensive interdisciplinary framework. The aim of this research was to develop and to introduce a practical methodology for evaluation of local and regional opportunities to promote CE.

Keywords Factors · Fuzzy cognitive map · Model reduction · Circular economy
Sustainability

1 Introduction

1.1 General Introduction and Motivation

The concept 'sustainability' is literally about maintaining of human existence. The importance of the topic of sustainability has become increasingly significant over the last fifty years due to an intensive population growth with massive per capita

A. Buruzs (✉)
Department of Environmental Engineering, Széchenyi István University, Győr, Hungary
e-mail: buruzs@sze.hu

M. F. Hatwágner
Department of Information Technology, Széchenyi István University, Győr, Hungary
e-mail: miklos.hatwagner@sze.hu

L. T. Kóczy
Department of Automation, Széchenyi István University, Győr, Hungary
e-mail: koczy@sze.hu; koczy@tmit.bme.hu

L. T. Kóczy
Department of Telecommunications and Media Informatics, Budapest University of Technology
and Economics, Budapest, Hungary

© Springer Nature Switzerland AG 2019
M. E. Cornejo et al. (eds.), *Trends in Mathematics and Computational Intelligence*, Studies in Computational Intelligence 796,
https://doi.org/10.1007/978-3-030-00485-9_22

resource consumption on a finite planet. Above all, it is necessary to clarify that sustainability is not an exact discipline such as, for example, mathematics. Rather it is a multidisciplinary science incorporating nearly all of human knowledge in approximately equal parts and with more or less equal importance. In sustainability, the combination of different disciplines appears, from the natural science, through social sciences to engineering sciences, including politics. The reasons for this is that sustainability is about sustaining human existence which requires many information and comprehensive knowledge to be maintained.

The term 'circular economy' is quickly attracted attention as a way of separating growth from resource limitations. It opens the way to comply the outlook for growth and economic participation with that of environmental awareness and equity. The concept of CE is to an increasing extent treated as a solution to series of challenges such as waste generation, resource scarcity and sustaining economic benefits [4]. In the last few years, CE is receiving increasing attention worldwide as a way to overcome the current production and consumption model based on continuous growth and increasing resource throughput. By promoting the adoption of closing-the-loop production patterns within an economic system CE aims to increase the efficiency of resource use, with special focus on urban and industrial waste, to achieve a better balance and harmony between economy, environment and society [5]. The circular economy provides a framework to challenge and guide us to rethink and redesign our future.

Such redesigns and decision problems are usually characterized by numerous issues or concepts interrelated in a complex way. Formulating a quantitative mathematical model for such systems and frameworks as circular economy may be difficult or impossible due to lack of numerical data and dependence on imprecise verbal expressions. An FCM is able to represent unstructured knowledge through causalities expressed in imprecise terms and simulate the operation of the system.

1.2 Structure of the Paper

The structure of the paper is as followings. In the 1 paragraph, section C, the literature review is provided. In Chap. 2, an overview on the applied method, namely the Fuzzy Cognitive Maps is given. In Chap. 3, the results of the complex model's simulation are introduced.

At the end of the paper, the authors concluded that for carrying out the proposed assessment methodology, expert knowledge is needed to ensure the reliability of the results obtained. The future research intentions are summarized in Chap. 5.

1.3 Literature Review

Almost all existing techniques evaluate resource use based on their burden relative to value, while the central point of Circular Economy (CE) is to create value through material retention. The existing burden-orientated techniques are therefore unsuitable for guiding managers in relation to CE objectives [1]. Programmes and policies for a CE are becoming key to regional and international plans for creating sustainable scenarios. Framed as a technologically driven and economically profitable vision of continued growth in a resource-scarce world, the CE is taken up by the European Commission and global business leaders alike [2].

Sustainability aims at addressing environmental and socio-economic issues in the long term. In general, the relevant literature on sustainability has focused mainly on the environmental issues, whereas, more recently, a CE is proposed as one of the latest concepts for addressing both the environmental and socio-economic issues. A CE aims at transforming waste into resources and on bridging production and consumption activities; however, there is still limited research focusing on these aspects [3].

1.4 Preliminaries of the Research

In order to model the problem under investigation, a method based on of Fuzzy Cognitive Maps (FCM) [6–8] was selected. This method is capable of simulating the operation of a model as long as the input data sets are available that include the factors with significant effects on the system and also the historical time series of these factors, which together allow the representation of the features of factors describing the operation of a systems. The authors' intention was to find out how to build up a sustainable waste and material management system.

2 Methodology Applied

The research is divided into two main phases. In the first one, the significant 33 factors (Table 1) of the waste management systems were determined [9]. The time series of these elements were developed based on the results of a text mining method [10]. Next, model calculations were made on the basis of fuzzy graph structure [11–13].

Table 1 The identified factors of the system

Main factor	Sub-factor	CID	Main factor	Sub-factor	CID
Technology (C1)	Engineering knowledge	C1.1	Society (C4)	Public opinion	C4.1
	Technological system and its coherence	C1.2		Public health	C4.2
	Local geographical and infrastructural conditions	C1.3		Political and power factors	C4.3
	Technical requirements in the EU and national policy	C1.4		Education	C4.4
	Technical level of equipment	C1.5		Culture	C4.5
Environment (C2)	Impact on environmental elements	C2.1	Law (C5)	Social environment	C4.6
	Waste recovery	C2.2		Employment	C4.7
	Geographical factor	C2.3		Monitoring and sanctioning	C5.1
	Resource use	C2.4		Internal and external legal coherence (domestic law)	C5.2
	Wildlife (social acceptance)	C2.5		General waste management regulation in the EU	C5.3
	Environmental feedback	C2.6		Policy strategy and method of implementation	C5.4
Economy (C3)	Composition and income level of the population	C3.1	Institution (C6)	Publicity, transparency (data management)	C6.1
	Changes in public service fees	C3.2		Elimination of duplicate authority	C6.2

(continued)

Table 1 (continued)

Main factor	Sub-factor	CID	Main factor	Sub-factor	CID
	Depreciation and resource development	C3.3		Fast and flexible administration	C6.3
	Economic interest of operators	C3.4		Cooperation among institutions	C6.4
	Financing	C3.5		Improvement of professional standards	C6.5
	Structure of industry	C3.6			

2.1 Fuzzy Cognitive Map

Based on the gathered data, the authors constructed the connection matrix. FCM consists of nodes and weighted arcs, which are graphically illustrated as a signed weighted graph with feedback. Nodes of the graph stand for the concepts describing behavioral characteristics of the system. Signed weighted arcs represent the causal relationships that exist among concepts and interconnect them. The relationships between concepts are described using a degree of causality. Experts describe this degree of influence using linguistic variables for every weight; so weight for any interconnection can range from $[-1, 1]$.

The degree of causal relationship between different factors of the FCM can have either positive or negative sign and values of weights express the degree of the causal relationship. Linkages between concepts express the influence one concept on another.

3 Results

The generated connection matrix contains 1056 (33*32) connection. Since the representation and interpretation of such a complex model is rather difficult, only the most important connections are represented in Fig. 1 with the help alpha-cuts.

With the help of α-cuts, the above graph clearly reveals that there are elements of the system which are very important – whether positive or negative effect – on the sustainability of the system's operation. It can also be seen which are of great importance on the other factors (factors with 2–4 outward impact). There are also factors which accept greatest impact (impact from 2–4 factors).

Accordingly, factors with outward impact are:

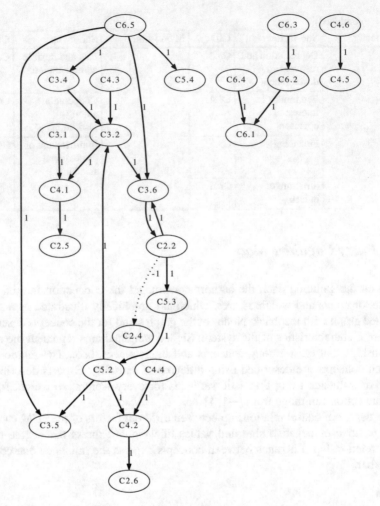

Fig. 1 The most important connection (−1 and 1) of the factors are represented with the help alpha-cuts

- 'Improvement of professional standards' (the greatest effect on four factors), followed by
- 'Waste recovery' and 'Internal and external legal coherence' (effect on three factors), and finally
- 'Changes in public service fees' close the line with influence on two factors.

  The most influence accepted by

- 'Structure of industry' (impact by 3 factors), and

  'Public opinion', 'Public health', 'Financing', and 'Changes in public service fees' (impact by 3 factors).

Accordingly, factors with outward impact are:

- 'Improvement of professional standards' (the greatest effect on four factors), followed by
- 'Waste recovery' and 'Internal and external legal coherence' (effect on three factors), and finally
- 'Changes in public service fees' close the line with influence on two factors.

The most influence accepted by

- 'Structure of industry' (impact by 3 factors), and

'Public opinion', 'Public health', 'Financing', and 'Changes in public service fees' (impact by 3 factors).

3.1 Model Reduction

During the simulation of this complex model, the authors proposed a new method to reduce the extent of the model, where three different distance definitions were introduced. The essence of this method is to create clusters of factors and using these clusters to develop new reduced models. Thus, reducing the number of factors, the model is more easily understandable and realistic.

Taking into account the special characteristics of the studied engineering problem, the number of clusters in the reduced model should be around six (based on the expert consensus [9]). Or, at least much less than the number of the factors in the detailed model.

The authors have made several attempts to find the right reduced model in order to determine the right clusters. On the one hand, if only some of the factors are merged, to achieve the goal of unification is not possible. On the other hand, if there are too many factors to be merged then the interpretation of the combined factors is difficult or almost impossible. The intention was to reduce the original matrix by a matrix with around 15 clusters.

The above mentioned three methods are different only in the metrics representing the similarity between the factors. This approach may be considered as a strong generalization of the state reduction procedure of sequential circuits or finite state machines [14, 15]. The essence of the methods is to create clusters from the factors and apply these clusters as the new factors in the new, simplified model. The base for binding these clusters is the use of a tolerance relation (a reflexive and symmetric, but non-transitive relation) [16, 17].

The authors have chosen matrices, each including 15 clusters. Tables 2, 3 and 4 introduce the clusters in the reduced connection matrix, using different metrics.

Tables 2, 3 and 4 show that there are overlapping between the clusters. Accordingly, a factor is listed 1 to 15 times in the clusters in the new models. The role of the factors in the clusters is introduced in Table 5.

Table 2 The clusters in the matrix produced by Metric A

1	C3.1+C3.2+C3.3+C3.4+C3.5+C2.5+C1.3+C6.1+C6.2+C6.3+C6.4+C6.5+C4.3+C4.4+C4.5+C4.6
2	C3.3+C3.4+C3.6+C2.3+C2.5+C1.3+C6.2+C6.3+C6.4+C4.3+C4.5
3	C3.1+C2.1+C2.5+C1.3+C6.1+C6.2+C6.4+C4.3+C4.7
4	C3.3+C3.6+C2.2+C6.2+C6.3+C6.4
5	C3.3+C3.4+C3.5+C2.3+C2.5+C1.1+C1.3+C6.2+C6.3+C6.4+C6.5+C4.3+C4.5
6	C3.1+C3.2+C2.4+C6.2+C6.3+C4.3
7	C3.1+C3.2+C3.3+C3.5+C2.5+C2.6+C1.3+C6.1+C6.2+C6.4+C6.5+C4.3
8	C3.3+C3.5+C2.3+C5.1+C5.3+C1.1+C1.2+C1.4+C1.5+C6.3+C6.4+C4.3+C4.5
9	C3.3+C3.5+C2.5+C5.2+C1.1+C1.3+C6.1+C6.2+C6.4+C6.5+C4.2+C4.3+C4.4+C4.5+C4.6
10	C3.3+C3.5+C2.3+C5.3+C5.4+C1.1+C1.4+C1.5+C6.2+C6.4+C6.5+C4.3+C4.5
11	C3.2+C3.3+C3.4+C3.5+C2.5+C1.1+C1.3+C6.1+C6.2+C6.3+C6.4+C6.5+C4.2+C4.3+C4.4+C4.5+C4.6
12	C3.1+C3.2+C3.3+C3.4+C3.5+C1.2+C1.5+C6.1+C6.2+C6.3+C6.4+C6.5+C4.3+C4.4+C4.5+C4.6
13	C3.2+C3.3+C3.4+C3.5+C1.1+C1.2+C1.4+C1.5+C6.1+C6.2+C6.3+C6.4+C6.5+C4.2+C4.3+C4.4+C4.5+C4.6
14	C3.3+C3.5+C5.1+C5.2+C1.1+C1.4+C1.5+C6.4+C4.1+C4.2+C4.3+C4.4+C4.5
15	C3.1+C3.3+C3.5+C2.5+C2.6+C1.3+C6.1+C6.2+C6.4+C4.3+C4.7

Table 3 The clusters in the matrix produced by Metric B

1	C3.1+C3.2+C3.3+C3.4+C3.5+C5.2+C5.3+C5.4+C1.1+C1.2+C1.4+C6.4+C4.6
2	C3.2+C3.3+C3.4+C3.5+C3.6+C5.3+C5.4+C1.1+C1.2+C1.4+C1.5
3	C3.1+C3.5+C2.1+C2.3+C2.5+C5.2+C5.3+C5.4+C1.1+C1.2+C1.3+C4.4+C4.5+C4.6
4	C3.2+C3.3+C3.5+C3.6+C2.2+C2.5+C1.1+C1.2+C1.5
5	C3.1+C3.3+C3.5+C2.3+C2.5+C5.2+C5.3+C5.4+C1.1+C1.2+C1.3+C1.4+C6.1+C6.3+C6.4+C6.5+C4.4+C4.5+C4.6
6	C3.5+C3.6+C2.3+C2.4+C2.5+C5.3+C1.1+C4.4
7	C3.1+C3.2+C3.3+C3.5+C2.5+C5.2+C5.3+C5.4+C1.1+C1.2+C1.4+C6.4+C4.6
8	C3.1+C3.4+C3.5+C2.6+C5.2+C5.3+C5.4+C1.1+C1.4+C6.1+C4.4+C4.5+C4.6
9	C3.3+C3.5+C2.2+C2.3+C2.5+C5.1+C1.1+C1.2+C1.5+C6.5+C4.4
10	C3.1+C3.3+C3.4+C3.5+C5.2+C5.3+C5.4+C1.1+C1.2+C1.4+C6.1+C6.4+C6.5+C4.4+C4.5+C4.6
11	C3.1+C3.3+C3.4+C3.5+C5.4+C1.1+C1.2+C1.4+C6.1+C6.2+C6.4+C6.5+C4.3+C4.4+C4.6
12	C3.1+C3.3+C2.3+C2.5+C5.2+C5.3+C5.4+C1.1+C1.2+C1.3+C1.4+C6.1+C6.3+C4.1+C4.2+C4.4+C4.5+C4.6
13	C3.1+C3.3+C2.3+C2.5+C5.2+C5.3+C5.4+C1.1+C1.2+C1.3+C1.4+C6.1+C6.3+C6.4+C6.5+C4.2+C4.4+C4.5+C4.6
14	C3.1+C3.2+C3.3+C3.4+C3.5+C5.3+C5.4+C1.1+C1.2+C1.4+C6.4+C4.3+C4.6
15	C3.1+C3.3+C2.3+C2.5+C5.2+C5.3+C1.1+C1.2+C1.3+C1.4+C6.1+C6.3+C6.4+C6.5+C4.2+C4.4+C4.5+C4.6+C4.7

Based on international experience, perhaps it is still surprising that the most important element of the system is 'Engineering knowledge'. It is followed by financial, technological and legal factors (C1.2, C3.3, C3.5, C1.4, and C5.3).

The authors concluded that Metric A to B and C shows a match of 75% in terms of the most common elements however B to C and vice versa shows a match of 94%. The authors also verified that 12 factors out of the most common 16 elements occur quite often in each cluster (max. 15, min. 4 times). In this sense, the metrics gave a very similar outcome.

As a result of the introduces fuzzy cognitive modelling techniques and the proposed new model reduction method it can be stated that the above listed factors can be of greatest importance on the promotion of the circular economy as a sustainable waste and material management system.

Table 4 The clusters in the matrix produced by Metric C

1	C3.1+C3.2+C3.3+C3.4+C3.5+C1.1+C1.4+C6.1+C6.4
2	C3.3+C3.4+C3.5+C3.6+C5.3+C5.4+C1.1+C1.2+C1.5
3	C3.1+C2.1+C2.5+C2.6+C4.5
4	C3.6+C2.2
5	C3.3+C3.5+C2.3+C2.5+C5.3+C5.4+C1.1+C1.4+C1.5+C6.4+C4.3+C4.4
6	C2.4+C2.5+C2.6
7	C3.1+C3.3+C3.5+C2.5+C1.1+C1.3+C1.4+C6.1+C6.3+C6.4+C4.3+C4.4
8	C3.1+C3.2+C2.6
9	C2.1+C2.3+C2.5+C5.1+C1.1+C1.3+C4.4+C4.5
10	C3.3+C3.5+C2.5+C5.2+C5.3+C5.4+C1.1+C1.4+C1.5+C6.4+C4.3+C4.4
11	C3.2+C3.3+C3.4+C3.5+C5.4+C1.1+C1.2+C1.4+C6.1+C6.4
12	C3.1+C3.3+C3.4+C3.5+C1.1+C1.4+C6.1+C6.2+C6.3+C6.4+C6.5+C4.3
13	C2.5+C5.1+C5.2+C5.3+C5.4+C1.1+C1.4+C1.5+C4.1+C4.2+C4.4+C4.5+C4.6
14	C3.1+C3.3+C2.5+C1.1+C1.3+C1.4+C6.1+C6.3+C6.4+C4.3+C4.4+C4.5+C4.6
15	C3.1+C2.5+C1.1+C1.3+C1.4+C6.1+C6.3+C6.4+C4.3+C4.4+C4.5+C4.6+C4.7

Since the indicated factors having the most significant effect on a system's sustainability receive the proper emphasis during the design and operation process, the effect of the other factors contributes also to the long-term management of the system.

The model formulated on the basis of the proposed method can be an example of how an environmentally and socially-economically mission can be done in a way to be able to provide a favourable solution from economic, legal and institutional point of view.

On the basis of the reduced model, the author make a proposal on what focus and what methods should be taken into account to set up a proper waste and material management system that meets the technical, legal and environmental requirements, and social expectations, as well as it can be operated economically in short and long term.

The 'Engineering knowledge' plays decisive role in design for proper technical, economic and environmental requirements. The factor 'Technological system and its coherence' has also major importance. The reason for this is that it is beneficial to reproduce the operation of the natural processes and ecosystems in the design of urban management activities. According to the previous principle, the waste and material management cycles should be organized and implemented in a closed way. In the development of the flow of material and waste cycles, the technology, its quality and its availability plays the biggest role. We should strive to create more closed systems through innovation.

It would be practical to develop a relatively self-sustaining system in this field. A decision support mechanism should be developed in order to find the most appropriate solution in specific time and place for (waste) materials in different quality and quantity.

Due to ensuring its promotion, it is necessary to make a decision in a short time about how and in what proportion used materials can return to the production cycle. Furthermore, to make decision about components that are unable to participate in the cycle.

Table 5 The role (frequency) of the factors in the new models

		Metric A	Metric B	Metric C
C1.1	Engineering knowledge	15	11	11
C1.2	Technological system and its coherence	12	2	2
C1.3	Local geographical and infrastructural conditions	5	4	4
C1.4	Technical requirements in the EU and national policy	11	9	9
C1.5	Technical level of equipment	3	4	4
C2.1	Impact on environmental elements	1	2	1
C2.2	Waste recovery	2	1	2
C2.3	Geographical factor	7	2	2
C2.4	Resource use	1	1	1
C2.5	Wildlife (social acceptance)	9	9	9
C2.6	Environmental feedback	1	3	3
C3.1	Composition and income level of the population	11	7	7
C3.2	Changes in public service fees	5	3	3
C3.3	Depreciation and resource development	12	8	8
C3.4	Economic interest of operators	6	4	4
C3.5	Financing	12	7	7
C3.6	Structure of industry	3	2	2
C4.1	Public opinion	1	1	1
C4.2	Public health	3	1	1
C4.3	Political and power factors	3	6	6
C4.4	Education	10	7	7
C4.5	Culture	7	5	5
C4.6	Social environment	11	3	3
C4.7	Employment	1	1	1
C5.1	Monitoring and sanctioning	1	3	2
C5.2	Internal and external legal coherence (domestic law)	9	2	2
C5.3	General waste management regulation in the EU	12	4	4
C5.4	Policy strategy and method of implementation	11	5	5
C6.1	Publicity, transparency (data management)	7	6	6
C6.2	Elimination of duplicate authority	1	1	1
C6.3	Fast and flexible administration	4	4	4
C6.4	Cooperation among institutions	8	8	8
C6.5	Improvement of professional standards	6	1	1

4 Conclusions and Future Research

The intermeshing of disciplines from the natural sciences, social sciences, engineering and management is essential to addressing this complex problem. The study presented here possesses also this feature for future applications.

Six to eight professionals with extensive experience in their fields are needed to support the fuzzy cognitive map methodology. The work of the group of experts needed to be moderated by an environmental specialist who also helps to interpret the results. So, at this stage of the evaluation the expertise and experience is of great importance. The support of an IT staff member is also required who performs the simulations based on the in-put data and help in producing results.

Therefore, it can be concluded that for carrying out the proposed assessment methodology, expert knowledge is needed to ensure the reliability of the results obtained.

The authors' purpose is to continue the investigation to understand the deeper context of the circular economy and try to develop a refined model.

Acknowledgements The authors would like to thank to EFOP-3.6.1-16-2016-00017 1 'Internationalisation, initiatives to establish a new source of researchers and graduates, and development of knowledge and technological transfer as instruments of intelligent specialisations at Széchenyi István University' and EFOP-3.6.2-16-2017-00015 "HU- MATHS – IN – Intensification of the activity of the Hungarian Industrial Innovation Service Network'for the support of the research.

References

1. Franklin-Johnson, E., Figge, F., Canning, L.: Resource duration as a managerial indicator for Circular Economy performance. J. Clean. Prod. **133**(1), 589–598 (2016)
2. Hobson, K., Lynch, N.: Diversifying and de-growing the circular economy: radical social transformation in a resource-scarce world. In: Futures, vol. 82, pp. 15–25, September 2016
3. Witjes, S., Lozano, R.: Towards a more circular economy: proposing a framework linking sustainable public procurement and sustainable business models. In: Resources, Conservation and Recycling, vol. 112, pp. 37–44, September 2016
4. Lieder, M., Rashid, A.: Towards circular economy implementation: a comprehensive review in context of manufacturing industry. J. Clean. Prod. **115**(1), 36–51 (2016)
5. Ghisellini, P., Cialani, C., Ulgiati, S.: A review on circular economy: the expected transition to a balanced interplay of environmental and economic systems. J. Clean. Prod. **114**(15), 11–32 (2016)
6. Zadeh, L.A.: Fuzzy sets. Inf. Control **8**(3), 338–353 (1955)
7. Axelrod, R.: Structure of Decision: The Cognitive Map of Political Elites. Princeton University Press (1975)
8. Kosko, B.: Fuzzy cognitive maps. Int. J. of Man-Mach. Stud. **24**(1), 55–75 (1985)
9. Buruzs, A., Hatwágner, M.F., Torma, A., Kóczy, L.T.: Expert based system design for integrated waste management. Int. J. Environ. Ecol. Geol. Geophys. Eng. **8**(12) (2014)
10. Buruzs, A., Hatwágner, M.F., Torma, A., Kóczy, L.T.: Retrospective reconstruction of time series data for integrated waste management. Int. J. Soc. Educ. Econ. Manag. Eng. **8**(12) (2014)

11. Buruzs, A., Hatwágner, M.F., Kóczy, L.T.: Expert-based method of integrated waste management systems for developing fuzzy cognitive map. In: Taher, A.A., Quanmin, Z. (Eds.), Complex System Modelling and Control Through Intelligent Soft Computations. Springer International Publishing Switzerland, pp. 111–137. Studies in Fuzziness and Soft Computing, vol. 319 (2015)
12. Buruzs, A., Hatwágner, M.F., Földesi, P., Kóczy, L.T.: Fuzzy cognitive maps applied in integrated waste management systems. In: IEEE HNICEM-ISCIII. Paper 63. 4 p (2014)
13. Buruzs, A., Hatwágner, M.F., Kóczy, L.T.: Using fuzzy cognitive maps approach to identify integrated waste management system characteristics. In: Baranyi, P. (Ed.), 5th IEEE International Conference on Cognitive Infocommunications: CogInfoCom 2014, IEEE Hungary Section, pp. 141–147 (2014). ISBN:978-1-4799-7279-1
14. Hatwágner, F.M., Buruzs, A., Földesi, P., Kóczy, L.T.: A new state reduction approach for fuzzy cognitive map with case studies for waste management systems. In: Somnuk, P.A., Thien, W.A. (Eds.), Computational Intelligence in Information Systems: Proceeding of the Fourth INNS Symposia Series on Computational Intelligence in Information Systems (INNS-CIIS 2014). 362p. Springer International Publishing Switzerland, Cham, pp. 119–127 (2015). Advances in Intelligent Systems and Computing; 331 (2014). (ISBN:978-3-319-13152-8; 978-3-319-13153-5)
15. Papageorgiou, E.I., Hatwagner, M.F., Buruzs, A., Kóczy, L.T.: A concept reduction approach for fuzzy cognitive map models in decision making and management. submitted to J. Neurocomputing
16. Das, M., Chakraborty, M.K., Ghoshal, T.K.: Fuzzy tolerance relation, fuzzy tolerance space and basis. Fuzzy Sets Syst. **97**(3), 361–369 (1998)
17. Klir, G.J., Folger, T.A.: Fuzzy Sets, Uncertainty and Information. Prentice Hall (1987)

Author Index

© Springer Nature Switzerland AG 2019
M. E. Cornejo et al. (eds.), *Trends in Mathematics and Computational
Intelligence*, Studies in Computational Intelligence 796,
https://doi.org/10.1007/978-3-030-00485-9

203